Sports Geek

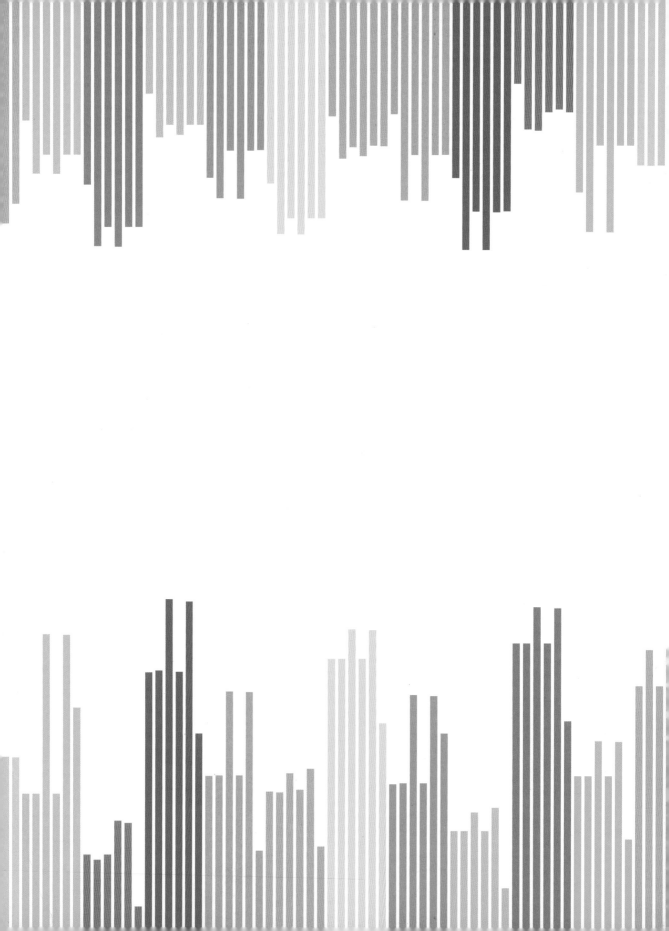

Sports Geek

A visual tour of sporting myths,
debate and data

Rob Minto

B L O O M S B U R Y
LONDON • OXFORD • NEW YORK • NEW DELHI • SYDNEY

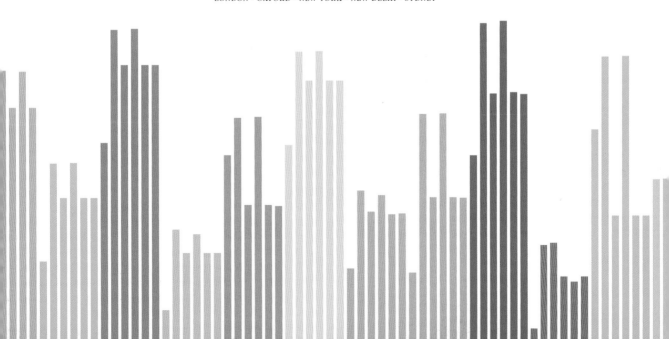

Bloomsbury Sport
An imprint of Bloomsbury Publishing Plc

50 Bedford Square
London
WC1B 3DP
UK

1385 Broadway
New York
NY 10018
USA

www.bloomsbury.com

BLOOMSBURY and the Diana logo are trademarks of Bloomsbury Publishing Plc

First published in 2016
© Rob Minto 2016
Data visualization by Beutler Ink

British Library Cataloguing-in-Publication Data
A catalogue record for this book is available from the British Library.

Library of Congress Cataloguing-in-Publication data has been applied for.

ISBN: Hardback: 978-1-4729-2749-1
ePDF: 978-1-4729-2748-4
ePub: 978-1-4729-2747-7

2 4 6 8 10 9 7 5 3 1

Designed by Carrdesignstudio.com
Printed in China by Hong Hing Printing Off-Set Co., Ltd. Shenzhen, Guangdong

s every effort to ensure that the papers used in the manufacture
ble products made from wood grown in well-managed forests.
form to the environmental regulations of the country of origin.

and books visit www.bloomsbury.com. Here you will find extracts,
hcoming events and the option to sign up for our newsletters.

Contents

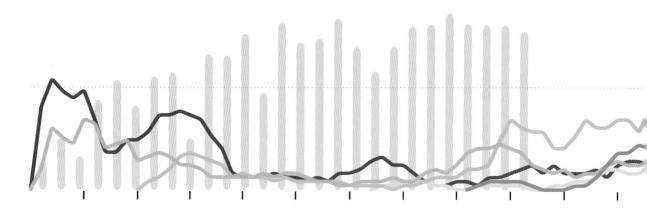

Introduction

I knew the moment I wanted to write about sport and data.

It was in the 1990s, when the BBC were showing the tennis at Queens, the tournament before Wimbledon. Pete Sampras was playing someone – I don't recall who – and the screen flashed up a new statistic.

It showed the number of break points Sampras had had on his opponent's serve, and how many he had won. It was something like one out of five – and even made it into a percentage: 20 per cent. (Leaving aside for now the usefulness of a percentage based on such a low number.)

It struck me that all those five break points came in one long game with several deuces. And Sampras still won that game. The break points lost didn't make any difference to the outcome of the match. If they had been spread over several games, you could make an argument of missed opportunities, but this told you next to nothing. This is a stupid stat, I thought. Surely next year they will come up with something better.

But year after year, the same break-point conversion statistic is floated on our screens. Why?

Because, paradoxically, sport and data often don't mix very well.

It's a paradox, because without numbers, sport doesn't work. We need to know the score. How fast the winner was. Whether it's a record. Sport attracts geeks as much as it attracts jocks; it's just that the jocks are louder.

Yet despite all the slew of data that Opta and others gather, despite the IBM trackers and new ways of counting tackles and points and relative speeds, we are no nearer understanding what works in sport.

Some so-called statistical insights tell us what we can already see: that some tennis players get a lot of cheap points on their serve, or that some football teams prefer to counter-attack than play possession.

Some stats tell us what we can't easily see – that winning football teams don't get more penalties than others, that longer golf drives don't make much difference.

And other statistics lead us down the garden path. Break points, meters covered, tackles made, yards passed: all can point to obvious but wrong conclusions if not used with context.

A great example of being misled by data is Sir Alex Ferguson, who admitted that he sold one of his best defenders, Jaap Stam, too early. Sir Alex saw figures that showed that Stam was making fewer tackles each season, so assumed it meant he was on the wane, and sold him to Italian club Lazio in 2001 for £16.5m. But Stam was making fewer tackles because he was playing 'smarter', not because he was performing poorly. A tackle is a last-ditch move in football. You want defenders shepherding, not tackling.

Football still hasn't learned much. We are now shown football statistics of how much ground players have covered, and those with the most metres on the clock are frequently praised. But where are they going? Are they simply chasing

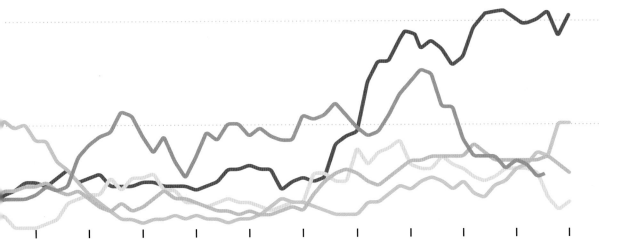

around pointlessly? When the ground-covered data for the England vs. Italy match at the World Cup in 2014 flashed up, the player who had covered the least ground (aside from the goalkeepers) was Andrea Pirlo. Pirlo, at age 35, was clearly the best player on the pitch. He didn't need to run excessively in the heat of Brazil to be in the right place at the right time.

So our assumptions and prejudices remain, compounded by new stats that we don't know how to use properly. Ex-players who are now pundits might not know the difference between causation and correlation, but it doesn't stop them (mis)using statistics.

Yet at the same time, there is amazing work being done in the field of sports statistics. There are brilliant books, like *The Numbers Game*, and *Soccernomics*. There is the MIT Sloan sports statistics conference, where ground-breaking ideas are put forward. It's a very exciting area.

Equally, behind the scenes, clubs are using data more and more. The irony is that they want to keep these insights for themselves, which means that the general public are given either a watered-down version, or kept in the dark altogether.

Which brings us to *Sports Geek*. This book isn't about new stats – most frequently it looks at data that has been around for years, and casts it in a new light. In some cases, I have taken public data that is in hard-to-read formats, or tucked away in thousands of web pages, then written programmes to compile it, and discovered new things. In other cases, the data is free and easy for anyone to access, it just isn't being interpreted very well.

Sports Geek is not about micro-level facets of the game – whether to pass or shoot. It is about the bigger picture of sport – about pay, nationalities, and corruption. It covers topics from film to failure, from drugs to demographics. And (hopefully) will entertain you along the way.

The overall message is not to obsess about data, but what the data highlights. Some chapters in this book show the limits of our knowledge, as the data is sketchy. How many golf courses are there in Asia? How many women footballers are there in the world? How many people actually watch the big sporting events on TV? In each case, the numbers are based on surveys, best-guesses and extrapolations. Yet many of these unreliable numbers are repeated with certainty in the press and by governing bodies.

Some chapters focus on great moment in sports, and break them down into their fascinating elements. Some tell the history of a sport through a single, key metric.

Elsewhere I have set out to create an argument, or to destroy a myth, or to explain sporting oddities. Is Usain Bolt slow at the 200m? Are Olympic medals easier to win? Are basketball teams losing on purpose?

The book covers over 20 sports. It is not necessarily meant to be read in a single sitting, although you are free to try. Instead, dip in and out. As sporting events come and go, it is your alternative reference book, a way to solve arguments, a companion to back you up when you know the commentator is talking total rubbish.

If you don't believe some of the conclusions, check out the sources. And if you simply disagree with something, that's OK. Remember: it's just sport.

World Cup Woes

Which country's fans suffer the most?

Something to complain about

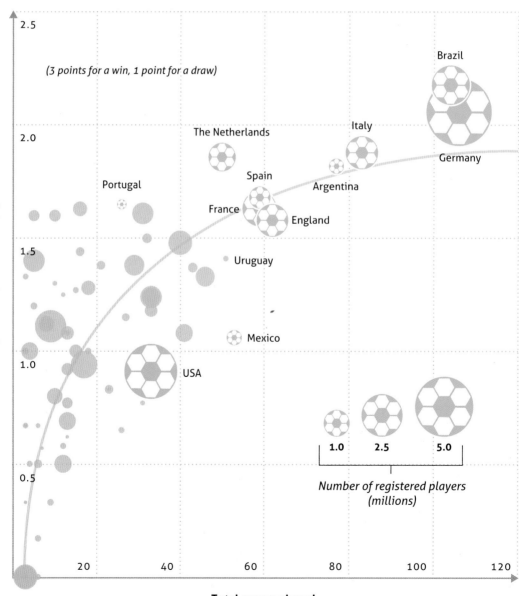

(3 points for a win, 1 point for a draw)

Average points per World Cup game

2.5

2.0

1.5

1.0

0.5

Brazil

The Netherlands

Italy

Germany

Portugal

Spain

Argentina

France

England

Uruguay

Mexico

USA

1.0   2.5   5.0

Number of registered players (millions)

20   40   60   80   100   120

Total games played

England fans tend to think that they are hard done by when it comes to World Cups. Despite the deep angst caused by dodgy or disallowed goals, usually against Argentina or Germany, they should get over it. When it comes to football's greatest prize, England aren't a bad team to support. The real pain lies elsewhere.

When we look at a country's overall World Cup record in terms of total games played, won, tied and lost, one country stands out for all the wrong reasons: Mexico.

No country has lost more games at the World Cup than Mexico – with 25 losses in 53 games, that's just about 1 in every 2 games played. Other countries have a worse losing record in percentage terms, but most of these are nations that are happy just to have qualified. The only other teams with a worse losing record that have played a comparable number of games are the US (19 losses in 33 games), Bulgaria (15 in 26) and South Korea (17 in 31). Switzerland and the Czech Republic are close, with 16 losses in 33 games. Put another way, Germany has played twice the number of games as Mexico, with 106, but lost 20 to Mexico's 25.

It's not as if Mexico plays all-out, boom-or-bust football, either – it ties lots of games as well. Mexico has won 14 World Cup games, and drawn 14. That puts it into the group of countries that have drawn as many or more games at the World Cup than they have won; and all of those have played significantly fewer games than Mexico. Nor can Mexico take any solace in being a small country. According to Fifa, it has more registered players than Argentina.

The chart shows points per game (3 for a win, 1 for a draw) at all World Cups, compared to total matches played. The two countries that stand out from the trend line are Mexico and at the top, the Netherlands. The different bubble sizes show the number of registered players in each country. Up the top end of the chart are Brazil and Germany, with the group of other World Cup winners trailing in their wake.

The team with the best record never to win the trophy is the Netherlands – it has a better win percentage (54 per cent) than Spain, England and France (all one-time winners) from a high number of games (50). On a points-per-game basis the country is behind only multiple winners Brazil, Germany and Italy. It is ahead of double winners Argentina.

Among the winners, Uruguay stand out as having a relatively poor overall record, with just over a third of matches won. That's owing to the country winning in the early years, 1930 and 1950, and doing little since bar a few fourth-place finishes.

For neutrals who like a result, Portugal is a good team to watch: in 26 World Cup matches, they have drawn only 4 times. (Hungary and Austria have similar records, but haven't qualified for a long time.)

Portugal is again of note when it comes to outperforming given the size of the country's talent pool: Fifa data show the country with just 132,000 registered players, but it has an average of 1.65 points per game, higher than France and England, which both have over a million players to choose from. Germany has the most registered players, with over 6.3 million. The US is second largest with 4m players, followed by Brazil with just over 2m. Other countries with over a million registered players include past winners Italy, as well as Japan, the Netherlands and South Africa. Clearly, a large number of registered players is no guarantee of success.

Nor is a large population in general: China has only qualified for one World Cup, scoring no goals and losing all three games, while India has never been to the finals. That's over a third of the world with little to no interest in football's biggest event – worth bearing in mind when talking of the global game.

And as for England's overall performance? It's very close to the chart trend line in terms of points and matches played.

So the next time England goes out on penalties or in other controversial circumstances, take heart from the fact that at least you aren't supporting Mexico – or the Netherlands, or the US, for that matter. In terms of World Cup history, things could be a lot worse.

How many women football players are there anyway?

In the rush to embrace the game, the numbers are getting inflated

'Let the women play in more feminine clothes like they do in volleyball. They could, for example, have tighter shorts.'

– Sepp Blatter, former Fifa President, 2004

'Fifa strives to promote gender equality and contribute to the empowerment of women worldwide.'

– Sepp Blatter, former Fifa President, 2013

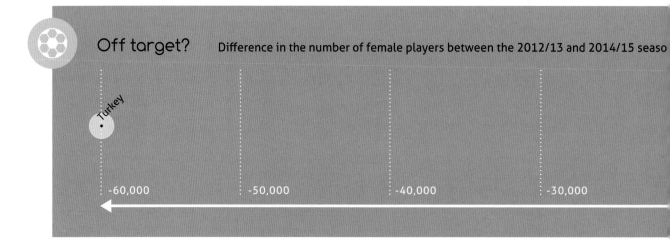

Off target? Difference in the number of female players between the 2012/13 and 2014/15 seaso

Turkey

-60,000 -50,000 -40,000 -30,000

Women's football has a credibility problem. Comments like those made by Sepp Blatter hardly help. The real problem though, is not on the pitch – anyone watching the World Cup in 2015 would agree that the game is fast, exciting and increasingly skilful. The problem is with the numbers.

In the rush to promote the game, big claims are made for the number of women playing, from four million to 30m. So how many women are actually playing the game?

Firstly, the football governing bodies count (or estimate) both registered and unregistered players. We will restrict our analysis to registered players, which should be easier to work out. However, this is where the problems start. Let's take Fifa, the overall world football governing body, and Uefa, who run European football.

Fifa's 2014 women's football survey lists nearly 4.8m registered women players, of which 2.25m come from the US

and Canada, and 2.1m come from Europe. But Uefa itself says there were 1.2m registered women players in Europe in the 2014–15 season.

That's a difference of just under 900,000 players. What's going on?

Let's track back. The source of the data isn't Fifa or Uefa. Neither organisation goes round football pitches in Albania or Iceland armed with a clipboard counting players. The data is supplied from the member associations. And they rely on clubs, schools and other groups to provide them with that data.

Some years the data look accurate, other years less so. For instance, the Austrian Football Association told Uefa that in 2014–15, it had 28,121 registered female players – very precise. But it also told Uefa that it had 20,000 players in 2013–14, 37,000 in 2012–13, and 17,000 in 2011–12, and (again) 17,000 in 2010–11.

Those numbers start to look a little less reliable. It seems unlikely that Austrian women footballers more than doubled in number in 2013 and then dropped back down again the following year.

Any time we see the same figure for registered players as the previous year, there's a good chance that the data haven't been collected. Uefa provides information for 54 member countries, with numbers of female players going back six years. Within that, there are 33 instances of repeated numbers of registered female players. That means

The information is collected directly from the national associations, as is Fifa's data. However, the number that you see in the Uefa publication does not take into account schoolgirls or those individuals who may retain a playing licence but no longer actively play football.

So is Fifa taking the same information as Uefa, and adding in potentially inactive players or schoolgirls? Fifa was contacted about the discrepancy, but at time of publication had not responded.

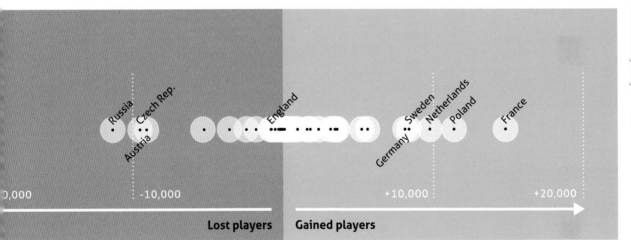

in the sample, 12 per cent of the information might be simply copied over from the year before.

It doesn't help that three of the bigger countries, France, Germany and England, failed to provide any numbers for the first three years (2009–12). And as the chart shows, the growth (or decline) in women players across countries is far from consistent.

Country associations also change their own rules and definitions. That is the explanation given for the drop in female players in Turkey, which has gone from 64,516 in 2014 to just 4,138 in 2015.

Still, that doesn't cover the huge difference in Fifa and Uefa data. So we asked both to clarify. Uefa said its definition of a registered player is someone 'whom the association has a record of actively playing football on a regular basis'.

Uefa also said:

It looks like Fifa has been overstating the number of women players for a long time. In its 'Big Count' survey of 2006, it claimed that there were 871,000 registered women players in Germany. Uefa, on the other hand, said in 2014 that there were 258,000 players in Germany. It's unlikely that half a million German women gave up football in seven years.

Let's just assume for a minute that Uefa's numbers are right and Fifa's are wrong, and extrapolate that figure worldwide. That would put the total registered players in women's football at 2.8m in 2014, more than a million lower than Fifa's headline number. It could well be a lot lower than that.

We may be on the cusp of something big for women's football, but using inflated numbers isn't the way to go about it.

Maradona's 10 seconds of genius

Argentina vs. England in Mexico '86 was no ordinary game of football

Argentina vs. England, the World Cup quarter-final, 22 June 1986. When Diego Maradona took control of the ball in his own half, and started his mazy run, you could sense something special about to happen. Watching it back now, even when he is still 40 yards from goal, the crowd noise is clearly deafening. The fans in the Estadio Azteca knew they might be about to witness a moment of magic.

There are three reasons why Maradona's second goal of the match is possibly the greatest ever.

First, there's the movement. It is, in and of itself, a perfect solo goal. Five players plus the keeper beaten; the speed and control of the football; at the starting point, in his own half, there seems no possibility of a goal – and yet it happened. Peter Beardsley, Peter Reid, Terry Butcher (twice) and Terry Fenwick got close to him, but could not get the ball. England goalkeeper Peter Shilton was left sprawling by a fleeting twist of a dummy before the ball – and Maradona – went past. He takes the last tackle, knowing it will put him down, but with the ball already dispatched a split second earlier into the net.

Second, there is the context of the match. Maradona had just scored the controversial 'Hand of God' goal, less than five minutes earlier, out-jumping Shilton and guiding the ball into the net with his fist. Whether the England players were distracted, still thinking about the injustice or not is impossible to know. Maradona, however, didn't stop. He was on a roll.

Third, there's the wider context. Sport is often a proxy for war, a prism for releasing our nationalist feelings and antagonisms. But England and Argentina had actually fought a war, only four years previously. This was the first sporting conflict since the Falklands, and it was keenly felt by both countries. A World Cup quarter-final is always a big match, but this was of another level of fervour.

Watching the grainy footage from Mexico '86 now, it seems like Maradona is moving in a different dimension to all the other players, obeying different laws. The commentary has that tinny quality of long-distance broadcasts. The crowd sounds enormous and exotic, and slightly crazed. In fact, it was the last time England played a match in front of over 100,000.

It was voted the Goal of the Century in a 2002 Fifa poll, so who are we to disagree? Is it the best goal ever? Very few others come close.

 Pick your commentary:

ENGLISH Jimmy Hill and Barry Davies

[Hill] "Here's Maradona again."
[Davies] "He has Burruchaga to his left, and Valdano to his left, he doesn't – he won't need any of them! Oh! You have to say that's magnificent. There is no debate about that goal. That was just pure football genius. And the crowd in the Azteca Stadium stand to him. Inside one, away from another, and the coolness under pressure to play the ball home with the side of his foot. If the first was illegal, the second was one of the best goals we've seen in this championship."

SPANISH Victor Hugo Morales

"Maradona has the ball, two mark him, Maradona on the ball, goes on the right, the genius of world soccer, he can pass it to Burruchaga, it's still Maradona! Genius! Genius! Genius! Ta-ta-ta-ta-ta-ta-ta. Goooooooooooal! Goooooooooal! I want to cry! Holy God! Long live football! Goooooooaal! Diegoaaaaa Maradona! It's enough to make you cry, forgive me. Maradona, in an unforgettable run, in the best play of all time. Cosmic kite!"

An anatomy of the greatest goal

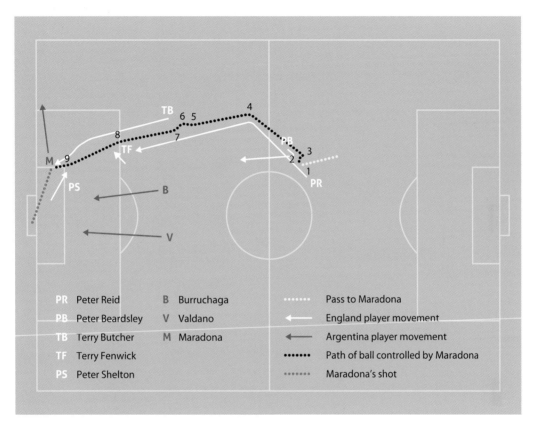

PR	Peter Reid	B	Burruchaga
PB	Peter Beardsley	V	Valdano
TB	Terry Butcher	M	Maradona
TF	Terry Fenwick		
PS	Peter Shelton		

- ········ Pass to Maradona
- ← England player movement
- ← Argentina player movement
- ●●●●●●● Path of ball controlled by Maradona
- ········ Maradona's shot

Steps	44
Players beaten	06
Touches of ball	10
Ground covered	60m
Time taken	10s
Argentina	2-1 England
Maradona 51', 55'	Lineker 81'

England
01 Peter Shelton (GK)(C)
02 Gary Stevens
03 Kenny Sansom
04 Glenn Hoddle
06 Terry Butcher
10 Gary Lineker
14 Terry Fenwick
16 Peter Reid (-69')
17 Trevor Steven (-74')
18 Steve Hodge
20 Peter Beardsley

Substitutes
11 Chris Waddle (+69')
19 John Barnes (+74')

Coach
Bobby Robson

Argentina
18 Nery Pumpido (GK)
02 Sergio Batista
05 Jose Brown
07 Jorge Burruchaga (-75')
09 Jose Cuciuffo
10 Diego Maradona (C)
11 Jorge Valdano
12 Hector Enrique
14 Ricardo Giusti
16 Julio Olarticoechea
19 Oscar Ruggeri

Substitutes
20 Carlos Tapia (+75')

Coach
Carlos Bilardo

The new champions, same as last time

European football's biggest problem is predictability

For 1999, see 1994. For 2003, see 1997. For 2010, see 2006. And 2015 looks much like 2005 did.

In those pairs of years, the same teams won the four big European football leagues: La Liga (Spain), Serie A (Italy), Bundesliga (Germany) and the Premier League (England).

Why should this matter? Statistically, it should be an aberration. Realistically, there are only a handful of teams in each league able to compete for the title. So the winners align from time to time.

It makes the football landscape that bit more predictable. Since Uefa started the money machine that is the Champions League in 1992, until 2015 Spain and Italy had only seen five clubs win each domestic league. Germany and England have seen six. Yet there is something of a paradox here: in the same period, the Champions League has been won by 13 teams, from seven different countries. It seems that the European club competition is a bit more of a lottery, or at least more open.

The more unlikely Champions League results are, however, withering away. Although teams from France, the Netherlands and Portugal have won in the past, the most recent 10 winners have all come from the four big European leagues.

The main reason is that the bigger leagues now get a handful of teams into the tournament, rather than just the winners of the domestic league. The first expansion was in 1998 to include runners-up. This was then increased in 2003 to the top four teams from Spain, Italy and England, and top three from Germany.

This means the top teams can retain players with the lure of European football with a high degree of certainty, and have greater experience playing in the event. And it means the inequality in resources and talent between the big teams and those outside Europe increases each season, with the obvious implications for predictability.

In the last 18 Champions Leagues, 10 of the winners have *not* been the winners the previous season of their domestic league, including the last four winners, which makes a nonsense of the name 'Champions League'. If the European clubs form a breakaway European league, as has been periodically threatened by the clubs and governing bodies, the surprise result won't just be scarce, it will become impossible.

Champions of Europe

The winners of Europe's four big domestic football leagues in the 22 years since the UEFA Champions League started

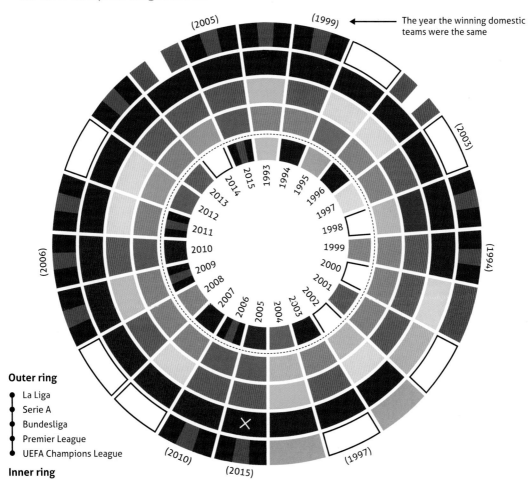

(2005)   (1999) ← The year the winning domestic teams were the same

Outer ring
- La Liga
- Serie A
- Bundesliga
- Premier League
- UEFA Champions League

Inner ring

England	Germany	Italy	Spain
Manchester United	Bayern Munich	Juventus	Barcelona
Arsenal	Borussia Dortmund	Milan	Real Madrid
Chelsea	Werder Bremen	Internationale	Atlético Madrid
Manchester City	FC Kaiserslautern	Lazio	Valencia
Blackburn Rovers	VfB Stuttgart	Roma	Deportivo La Coruña
	VfL Wolfsburg		

Other

Marseille	Ajax	Porto	Liverpool

Demography isn't destiny

Why aren't China and India any good at football?

There are two standard responses when the subject of China and/or India's football teams comes up. One is 'With all those people, you'd think they could find a half-decent team.' The other is: 'With that many people, they must be successful one day.'

Both ideas are fallacies. Here's why.

1) Population is a poor guide to performance

From Jamaica at sprinting to New Zealand at rugby to Norway at skiing, small populations can produce what seems like a disproportionate number of winners. The reasons are manifold: tradition, genetics, facilities, climate, wealth. And the opposite is also true: large populations don't guarantee a high level of performance across all sports. We might as well ask why Indonesia (the fourth largest country in the world, pop. 257m) isn't better at rugby.

2) Priorities

Do China or India care about football? The answer is: not enough. India cares far more about cricket. China has a decent interest in football, with more television coverage and an influx of foreign players from Brazil and elsewhere in the domestic league.

The interests of the Chinese people are not the same as those in power. The Chinese authorities have invested more heavily in individual sports to maximise Olympic medals, a strategy which paid off handsomely in Beijing 2008 with 51 golds and over 100 in total. Meanwhile, the domestic football league has been beset by corruption and match fixing. China

2 billion people, 22 players

Fifa ranking (end of year)   ● China   ● India

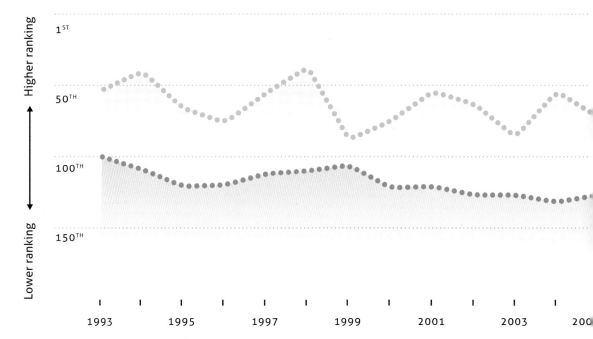

now has a 30 year plan to produce a top world-class team, but that's a long way off.

3) Talent pool

China and India might have over 1bn people each, but they don't actually have that many football players. According to Fifa, China has 711,000 registered players, and India has 385,000. This puts India 24th in the world, and China 12th. But when taken as a percentage of their overall populations, that works out at 0.05 per cent for China and 0.03 per cent for India. By contrast, Germany has 7.8 per cent of its population registered as players, and the US has 1.3 per cent.

Being a registered player in India or China is a very unusual thing to be. They are a tiny subsection of the population. That means fewer role models (on a local level, rather than in the media) and a talent pool that is insignificant nationally.

4) Urbanisation

You might think that cities are not conducive to playing football, but in fact, higher degrees of urbanisation lead to better football performance.

The latest UN urbanisation data for China and India put them on 52 per cent and 31 per cent. That will have gone up a bit, but both countries are still well behind the typical developed country level of around 70 to 80 per cent.

5) Where are the stars?

Name an Indian sports star. Ok, aside from Tendulkar? It's unlikely to be a footballer, that's for sure. Now name a Chinese sports star. Tennis player Li Na? Yao Ming of the NBA? Again, no footballers. If you want kids to take up a sport, they need to dream of being someone.

The Fifa world rankings have produced some unusual results over the years, but they are the best thing we have to go on. The interesting thing is that China and India are getting worse, not better, over time. The chart shows their year-end ranking over the last 22 years. The pattern is clear.

This is the rebuttal to the idea that 'one day' China and India will produce successful teams. When it comes to economics, demography is destiny. Not in football.

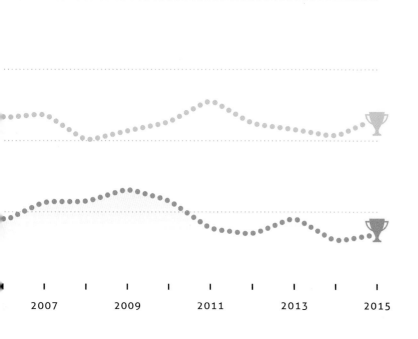

2007   2009   2011   2013   2015

Should better teams get more penalties?

Spot kicks aren't handed out for league position

Every Saturday there are several conversations that go a bit like this:

Interviewer: *What about the penalty decision late on?*
Manager: *Yes, I thought it should have been a penalty. Our player was clearly fouled, and the referee should have given it.*
Interviewer: *What about your opponents? They had an earlier decision go against them too.*
Manager: *I didn't see it.*

There's rarely a game of top-flight football without a controversial penalty decision, whether awarded or not given. And as penalties have a disproportionate impact on the outcome of football games,* it's not surprising that managers get so worked up about them. Their careers can change with one blow of a referee's whistle.

Naturally, managers are prone to bias when it comes to refereeing decisions. None more so than Jose Mourinho while he was Chelsea manager, who not only claimed that the decisions were unfair, but went further saying there was a reluctance amongst referees to give penalties to his

Premier League winners: penalised?

Penalties awarded to league winners

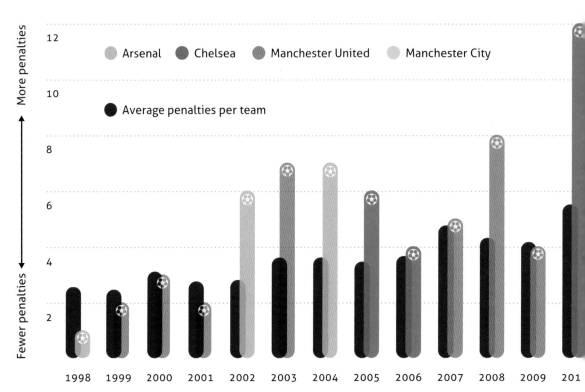

team. In October 2015, he said: 'Referees are afraid to give decisions to Chelsea ... be honest with us and give what you have to give.'

While it goes with the territory for managers to be biased in their assessment of decisions, it is quite another thing when a club makes purportedly statistical claims that they should have more penalties awarded.

Yet that is exactly what Chelsea FC did in 2015. In a statement^, the club said that at that point of the season, the number of penalties the club had been given was 'abnormally low'.

As well as lots of examples of when Chelsea felt a penalty should have been given, the club asserted that. 'Our position as clear league leaders and second-highest scorers suggests

we can't be labelled anything other than an attacking side, spending plenty of time in the opposition box.'

So let's look at the evidence. The numbers that Chelsea used only looked at their own penalties awarded. Statistically, it's known as sampling bias. Instead, we will look at the Premier League as a whole. And then throw in La Liga, Serie A and the Bundesliga for good measure.

The results are clear: there is only a very weak correlation between penalties over a season and a team's points. For the years of data available, the results are:

	Correlation of penalties to league position [1]	Correlation of penalties to points
Premier League (1998-2015)	-0.27	0.31
La Liga (2001-2015)	-0.27	0.34
Serie A (1998-2015)	-0.23	0.25
Bundesliga (2005-2015)	-0.24	0.26

[1] Correlation to position is negative as better teams have a lower number. Correlation scores of less than 0.5 are considered weak. Perfect correlation is 1 (or -1), perfect randomness is zero.

What about the league leaders? Are they (as Chelsea suggest) a special case?

Not really. This chart shows the penalties awarded to the Premier League winners each season, compared to the league average. Some years, the winners get a lot. Some years, not so many. There are plenty of examples of teams lower down the league that win a lot of penalties – partly because when it comes to a relegation fight, clubs attack just as much as when they are trying to win the league. For instance: Crystal Palace in 2004-05 were relegated from the Premier League, finishing in 18th place. Yet they topped the league with 12 penalties. In 2001 and 2015, the teams finishing first and last were given the same number of penalties.

Equally, there are a few examples of the league leaders getting the most penalties. The point is, it's not a given. Like most Chelsea appeals, apparently.

*See *Soccernomics* by Simon Kuper and Stefan Szymanski for a full analysis of how penalties influence games.

^ http://www.chelseafc.com/news/latest-news/2015/03/penalty-puzzle.html

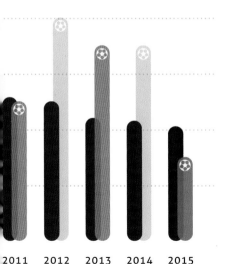

2011   2012   2013   2014   2015

European football's fallacy

Every year it gets more competitive, doesn't it?

Premier League managers maintain the same mantra each season: it's getting more competitive; every team can beat those at the top. And so on. Then came Leicester CIty, and blew apart the assumptions about money and sucess in 2015-2016. The club avoided relegation in 2015 by a whisker. It then defined the odds (famously 5000-1 at one point) to win the League. Surely that shows how truely competative football has become?

Maybe people mean different things by 'competitive'. One meaning is rising standards. Another is that there's a more level playing field, that teams are more evenly matched.

Rising standards are not our concern. They are too subjective. We are more interested in how to measure how evenly matched a league is. And there's one obvious way: the points each team accumulates over a season. As managers and pundits say, the table doesn't lie.

A competitive league should have a close cluster of points; a less competitive league would have a wide distribution, with teams racking up lots of points at the top, with the bottom clubs struggling.

It's tempting to point to one-sided matches, such as the number of 4–0 scorelines per season per league. But this is misleading: a team can be utterly dominant and still

Europe's big four leagues: more competitive?

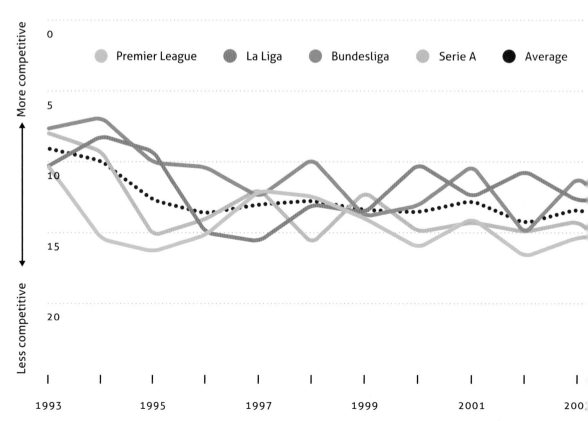

draw 0–0. Points accumulated over a season is admittedly imperfect, but a better guide to a team's relative strength to its peers.

The way to measure a distribution of numbers is standard deviation. Leaving aside how it is measured, a high standard deviation of a group of numbers shows a wide spread; a low standard deviation score shows a closer grouping.

To create a proxy for competitiveness, we have taken the standard deviation of each season's final table for the big four European leagues. Competitiveness goes up when the standard deviation goes down, and vice versa.

And the chart is clear: the trend in all four big football leagues of Europe – the Premier League (England), La Liga (Spain), Bundesliga (Germany) and Serie A (Italy) – is for competitiveness to decrease. The gap between teams is widening, not getting smaller. Spain especially has seen a fall in competitiveness over the last few seasons.

Of course, it only takes a glance at the winners to see that the pool of potential champions is small in every league, despite Leicester's unbelieveable season. But that isn't a measure of competitiveness either. Standard deviation is blind to which teams come out on top – it simply measures how tightly contested the league is, from top to bottom.

There was a slight pick-up in competitiveness in the 2014–15 season, but the overall trend in Europe is down, not up. The managers and pundits are simply wrong: the leagues are becoming less competitive.

Standard deviation of final points standings

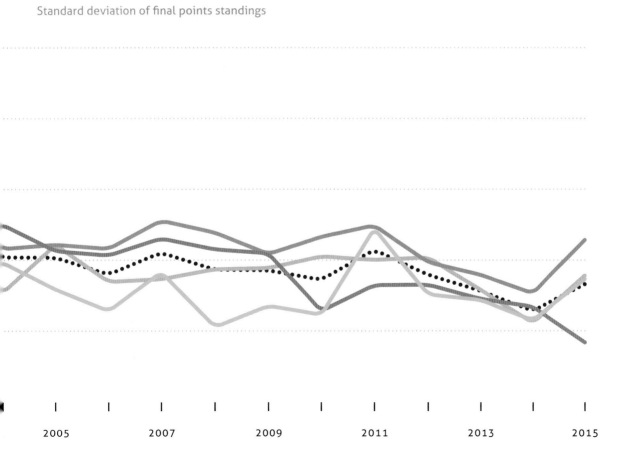

2005   2007   2009   2011   2013   2015

Dangerous duopolies

Which European football league is worst afflicted by the two-horse race?

History repeating

European leagues by concentration of winners

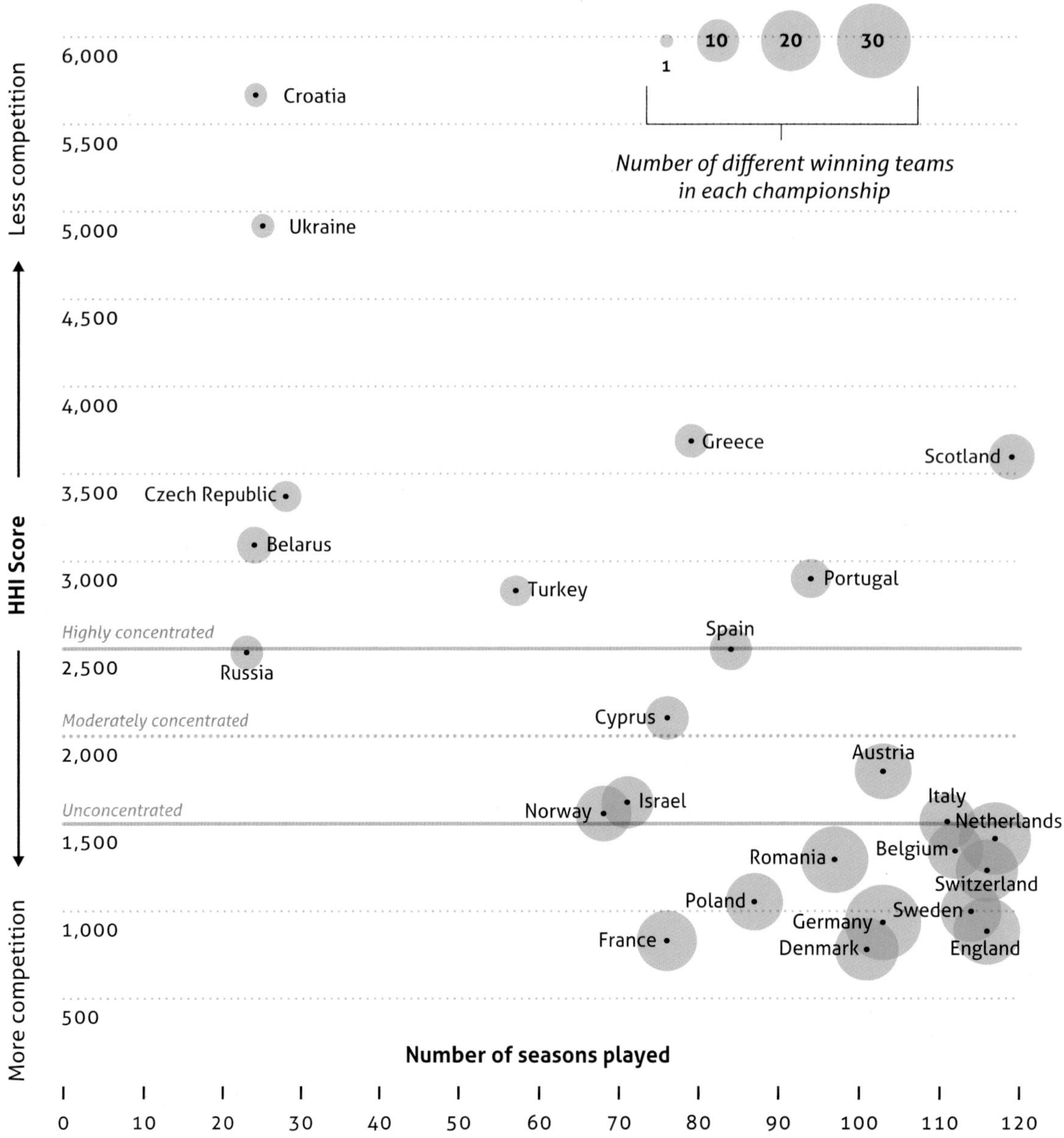
1
10
20
30
Number of different winning teams in each championship
Less competition
HHI Score
More competition
6,000
5,500
5,000
4,500
4,000
3,500
3,000
2,500
2,000
1,500
1,000
500
Highly concentrated
Moderately concentrated
Unconcentrated
Croatia
Ukraine
Greece
Scotland
Czech Republic
Belarus
Turkey
Portugal
Spain
Russia
Cyprus
Austria
Israel
Norway
Italy
Netherlands
Belgium
Romania
Switzerland
Poland
Sweden
Germany
France
Denmark
England
Number of seasons played
0
10
20
30
40
50
60
70
80
90
100
110
120

When Atlético Madrid won Spain's La Liga in 2014, it was a something of a sporting miracle. A team with a fraction of the financial clout of Real Madrid or Barcelona had played Moneyball, but, unlike their namesake Oakland Athletics, actually won. It also broke up one of the longest-running duopolies in Europe, going back ten years. Similarly, Leicester City's 2016 Premier League triumph put an end to two decades of just four different winners.

European football thrives on rivalries. Some are almost country-within-country battles, like Real Madrid and Barcelona. Others, like Celtic and Rangers (until recently), Fenerbahçe and Galatasaray, are only few miles apart in the same city. All are examples of the same problem: European football leagues dominated by two clubs. The rest don't get much of a look in.

While many English football fans complain that the Premier League is something of a closed shop, it's worth bearing in mind that in 1999, as Manchester United and Arsenal were locked in the middle of their period of domination, Manchester City were in the third tier division, having been relegated twice. They were champions just 12 seasons later. Money may play a large part in the club's revival, but it's still a change.

To measure how dominated a league is by few clubs, *Sports Geek* has borrowed a formula used to look at oligopolies (a term used to describe markets dominated by a few companies) and used it to analyse the top 25 European leagues, according to Uefa. The measure used is the Herfindahl-Hirschman Index (HHI*), and the maximum score is 10,000 for a monopoly, with zero for perfect competition (see explanation below).

If we look at each European league's roll-call of champions, there is a clear correlation between how many years a country's league has been going and its HHI score.

This reflects the fact that as you go further back, football success was far less determined by money and more clubs could compete.

Two leagues that were created in post-Soviet Europe – the Croatian and Ukrainian leagues – score very highly – indicating a lack of competition, with only a few clubs winning the league in their short history. Yet for Russia and Belarus, that short history has resulted in scores comparable to the far-longer established leagues of Turkey and Portugal. In terms of Europe's more prestigious leagues, Germany, England and France do well, with HHI scores that would be regarded as 'unconcentrated' in market terms and 'wide open' in commentators' parlance.

Italy would just fall into the moderately concentrated group with a score of 1,512. Spain are in this group too, but only just – another championship win for either Real Madrid or Barcelona would tip the country into the next level with a HHI of over 2,500. In 84 years of competition, only nine clubs have won the main club title in Spain.

The country that stands out for longevity *and* being uncompetitive is Scotland with a HHI score of 3,593. It has the longest-running duopoly in Europe, with either Celtic or Rangers winning the title every season since 1985. And with Rangers now clawing their way back from bankruptcy and demotion to the fourth tier of the Scottish Football League, Celtic look set to keep on winning.

Of course, some leagues are more long term monopolies than duopolies. Bayern Munich have won 25 out of 103 German football championships, more than double the next best team. Likewise, Anderlecht of Belgium, with 33 titles out of 112. But both leagues have quite low HHI scores as there is a greater number of other title-winning teams when Bayern or Anderlecht slip up. League domination comes in different statistical forms.

---

*Herfindahl-Hirschman Index – HHI

This is a widely-used measure of market concentration. To get the HHI score, take each company's market share as a percentage, square it, and add them up. In this case, we take each club's league championships as a percentage of total league seasons, square that, and add up the total.

If one team had won every league, the score would be 10,000. If a league existed where, over 100 seasons, a different team won every time, the score would be $(100*(1\textasciicircum 2))=100$.

The US Department of Justice uses the HHI to look at potential mergers. It says that unconcentrated markets have a HHI below 1,500; moderately concentrated markets have a HHI between 1,500 and 2,500; and highly concentrated markets have a HHI above 2,500.

The most complicated tournament in the world

How to make sense of the Europa League

It was once called the Uefa Cup, and it was a pure knock-out tournament. Teams qualified by doing well in their leagues, but not quite winning. Life was simple.

But then it got complicated. In what is a classic example of sporting mission creep, Uefa merged two other cups (the Cup Winners Cup in 1999 and InterToto Cup in 2009) with

104
Qualifying round

52+14
Qualifying round 2

33+25
Qualifying round 3

29+15
Play offs

This stage takes place in early July, when most of Europe is on summer holiday. The 104 teams that take part include the Gibraltar Cup winners, the San Marino league runners up, and the third-best team in the Faroe Islands. Three teams are included on the basis of fair play, regardless of performance.
Total matches: 104

Still July. The 52 teams that won in the previous round are now joined by 14 more from around Europe. Teams such as the Cypriot Cup winners, or the league runners up from Israel.
Total matches: 66

Late July, early August. The 33 winners from Q2 are now joined by 25 more from around Europe, with some of the bigger leagues taking part. The sixth-best teams from Spain, Germany and England are included.
Total matches: 58

Still in August, the play-offs feature the 29 winners from Q3, plus 15 teams that couldn't cut it in the more prestigious Champions League. Any team that has survived from Q1 by this point is doing very well, given the influx of bigger teams.
Total matches: 44

the Uefa Cup, and called it the Europa League. Along the way, in 2005, the competition took in the not-quite-so-good teams from the higher European competition, the Champions League.

Now the second-tier European club competition casts its net far and wide. As an example, the third-best team in the Armenian league is in the same competition as the cup winners from Spain.

There are now seven different entry points into the Europa League, from the first qualifying round to the round of 32. There are two-legged ties, a group stage, more two-leg matches, and eventually a final.

This chart tries to make sense of the tournament, which runs from June to May, an 11-month marathon that for some teams is more of a curse than a blessing.

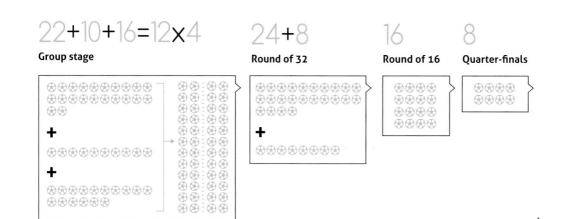

$$22+10+16=12\times4$$

Group stage

$$24+8$$

Round of 32

$$16$$

Round of 16

$$8$$

Quarter-finals

The group stage starts in September and runs until December. The 22 winners from the play-offs are again joined by an influx of Champions League rejects who didn't make it through their play-offs, and 16 others. The groups of four teams play each other home and away.
Total matches: 144

Phew, the group stage is over, and after a January break, the 24 teams that came first or second in their group move to the knockout stage. But what's this? Another eight Champions League failures join in, making 32.
Total matches: 32

No more surprises - the winners of the previous round play a two-leg match in March.
Total matches: 16

As you'd expect, the teams play home and away. It's April.
Total matches: 8

$$4$$

Semi-finals

$$2$$

Final

$$1$$

Winner

Down to the last four teams, who play in April and early May.
Total matches: 4

A one-off match, neutral venue. Late May.
Total matches 1

At last...

TOTALS
TEAMS: 192
MATCHES: 447

Fewest matches for a team to win: 9
Most matches for a team to win: 23

The sporting jamboree and corruption

Whether it's bids or regimes, corruption and sport have always gone together

When Oslo pulled out of bidding for the 2022 Winter Olympics in 2014, it was a watershed moment. The Norwegian parliament had voted against supporting the bid – which had been the favourite – mainly over issues of cost. Oslo joined the other withdrawn bids of Stockholm (Sweden), Krakow (Poland), Lviv (Ukraine), St Moritz (Switzerland) and Munich (Germany), all of which had been scrapped after votes or public discontent.

That left just Beijing in China and Almaty of Kazakhstan in the running. Neither were ideal hosts in terms of human rights. Nor are they democracies. After the exorbitant cost ($50bn) of Sochi 2014, and the allegations of corruption, the Winter Games was left to two authoritarian regimes. Beijing won, leaving many to wonder why the Winter Olympics was going to a place with no snow.

The football World Cup host selection process is under even greater scrutiny, following the indictment in May 2015

of many senior Fifa executives by the US Department of Justice. The decision to award Russia the 2018 World Cup and Qatar the 2022 edition seemed to confirm that only corrupt regimes with petrodollars to burn and a desire to strut the world stage need apply for the big jamboree events.

Are hosting the Olympics and football World Cup now just a dictators' sport? It's tempting to bemoan the new world order, but in reality, it was ever thus.

The 1988 Seoul Olympics may have marked South Korea's emergence economically, but the bidding process started at the end of military dictatorship. Moscow in 1980 was still under Communist one-party rule. The 1978 World Cup was hosted by Argentina, under the rule of a military junta, and (allegedly) was rife with match fixing, drugs and torture of dissidents. We can go back further: 1936 was Hitler's Olympic Games in Berlin. Mussolini made sure that Italy won the 1934 World Cup on home soil.

Big sporting events: getting more corrupt?

Transparency International's Corruption Perception Index

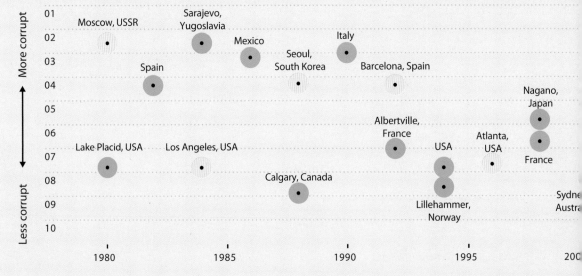

In 2013 Fifa secretary general Jerome Valcke stated: 'Less democracy is sometimes better for organising a World Cup.' Enough said.

Conversely, London 2012 and Tokyo 2020 are hardly Olympics under dictators. Brazil, although dogged with a corruption investigation, has hosted the 2014 World Cup and 2016 Olympics as a functioning democracy. Germany hosted the World Cup of 2006. The US and Sweden are looking at bids for future Olympics.

So how can we measure the honesty of hosts of sporting jamborees? GDP, cost per capita, and other measures are all interesting but don't reveal much about the hosts in terms of probity. Instead, *Sports Geek* has taken the well-regarded corruption index scores from Transparency International, a non-governmental organisation.

Caveat: the scores are taken for the nearest available year, with the earliest 1995. The year taken is when the event was staged, not bid for.

And the result? As the chart shows, the idea that increasingly only corrupt regimes win the big events doesn't hold true. Even Qatar has a corruption score of 6.9, higher (i.e. better) than both South Korea and Italy.

In fact, while we might assume Europe is the good guy in all this, Spain, Italy and Greece scored badly in the early editions of Transparency International's index, all below 5. Russia is the worst, at 2.7.

This of course doesn't factor in the possibility that supposedly honest countries have crooked bids. The Japanese Nagano Winter Games bid featured a rule-breaking party. London's 2012 bid faced allegations of corruption. Salt Lake City's 2002 bid was riddled with bribery. Even the Germany 2006 World Cup bid is under suspicion. It is tempting to conclude that when it comes to the big sporting events, it's safe to assume that everyone is breaking the rules. Some are just better at it.

NOTE:

Transparency International's Corruption Perception Index changed from being a score out of 10 to out of 100 in 2012. For the chart, the 2012 scores onwards have been divided by 10 for consistency.

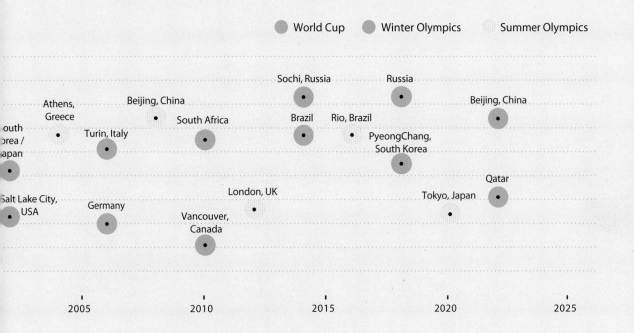

Home comforts

Is hosting the Olympics worth it? Yes, unless you are the United States

Medal spike: The Olympic host effect

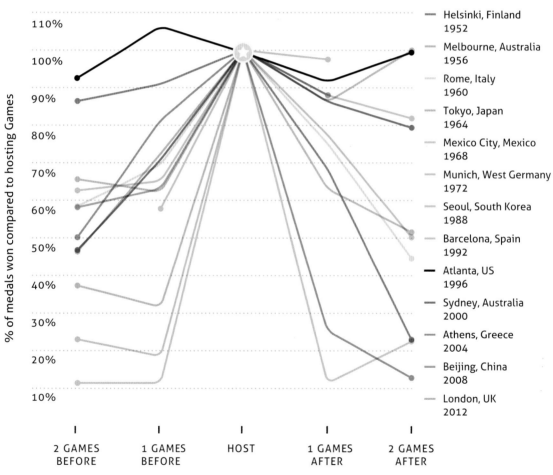

Helsinki, Finland 1952
Melbourne, Australia 1956
Rome, Italy 1960
Tokyo, Japan 1964
Mexico City, Mexico 1968
Munich, West Germany 1972
Seoul, South Korea 1988
Barcelona, Spain 1992
Atlanta, US 1996
Sydney, Australia 2000
Athens, Greece 2004
Beijing, China 2008
London, UK 2012

% of medals won compared to hosting Games

2 GAMES BEFORE   1 GAMES BEFORE   HOST   1 GAMES AFTER   2 GAMES AFTER

Hosting the Olympics is often likened to a big party: there's a lot of stressful preparation, it costs a hell of a lot, and afterwards you aren't sure if it was worth it. But that's the economics. In terms of winning medals and putting a country on the sporting map, it can really pay off. Perhaps there's a different party analogy: it's like your fifth birthday and you get to win pass-the-parcel several times. For sporting glory, hosting the Olympics is a better bet than the World Cup, which costs a lot too but only has one winner. In the Olympics, there are lots of medals to go round.

There have been many academic studies of the hosting effect of the Olympics, but *Sports Geek* is going to do something a bit simpler. Let's take the number of medals that a host country wins. Then compare it to the number of medals won in the two Games preceding hosting, and the two Games after, and show those as a percentage of the hosting total. In modern times, hosting has usually been awarded seven years prior, so the Games eight years before hosting should show the country's normal medal level. (This is for Summer Games only, as the Winter Olympics naturally favours countries with snow and a history of winter sports.)

This allows us to compare Olympic hosts relatively effectively, as every host country is given a score of 100. It's not perfect in that it doesn't take changes in GDP and population into account; nor does it factor in the number of competitors or countries. But this is not a scientific study. It just shows what actually happened.

The Olympics stopped for the two world wars, which meant hosting became more haphazard. Also, there were the boycotts of 1976, 1980 and 1984, which make things difficult. Moscow hosted in 1980 but the Soviet Union boycotted Los Angeles in 1984; the US boycotted the 1980 Games; and Canada hosted in 1976 but boycotted 1980 as well. That takes three series of medal results out of the data.

Still, looking at post-WWII Games only, we are left with 13 Olympic hosts and their medal progression. The Rio de Janeiro Games of 2016 couldn't be included in this book owing to publication deadlines. But given that Brazil won 15 medals in Beijing and 17 in London 2012, and looking at the average and median uplift from hosting, a total medal haul of around 25 to 30 would be my guess.

The chart shows a clear trend for hosts to win far more medals than they normally get – as we would expect. But there are a few interesting observations.

Every country that has hosted the Summer Games has seen a substantial boost, except for the US. It won 108 medals in 1992 compared to 101 in Atlanta in 1996. What went wrong?

Partly, this is because the US is such a big country with a strong sporting culture that it does well at every Olympics, regardless of venue. The host effect is less pronounced.

Also, as LA and Atlanta were so close together, it's hard to separate the legacy of LA from the hosting of Atlanta.

The post-Games effect is slightly bigger than the pre-Games effect. While a few hosts fall away in subsequent Games (Greece and Mexico in particular), Olympic hosts tend to do well in the next Games too. China, South Korea, Spain, West Germany, Australia (Sydney), and Japan have all seen better medal returns in the Games after hosting compared to the Games before. Canada, which we have not included due to the 1980 boycott, won 44 medals in 1984 in LA, a huge leap from the 11 it won as hosts in Montreal in 1976. In effect, LA was like a home Olympics for the Canadians; the country has never done as well since.

So what's behind the home-advantage at the Olympics?

One reason is investment: when a country wins the hosting process, it usually embarks on an extra-intensive programme of identifying and backing athletes who will peak at the Games. Every country tries to peak for the Olympics, but the hosts are especially driven in this regard.

Another boost may come from being used to the facilities and environment: you're used to the weather, you know the track. Certainly that has been shown to be a contributing factor in several studies of home advantage in various sports as well as the Olympics.

Other factors that help the hosts, however, may be less than honest. One is undetected doping. Again, this is impossible to prove, and countries other than the hosts may be using drugs to boost performance. There is bias. A 2003 paper* looking at Olympic home advantage found that it was significant in events that 'are either subjectively judged (gymnastics, boxing) or rely on subjective decisions (team games)'. Meanwhile, measured events such as athletics and weightlifting 'showed no home advantage'.

In other words, the referees and judges are biased, either nefariously or simply through the effect of crowd noise and pressure. Home referee bias has been identified in several other sports, the different ways of measuring success at the Olympics exposes the effect even more.

*'Modelling home advantage in the Summer Olympic Games', N J Balmer, A M Nevill, A M Williams, *Journal of Sports Science*, June 2003

Faster, higher, easier?

How the Olympics is getting less competitive

Each Olympic Games sees a clutch of new records. And we are often told by commentators and experts that there are rising standards across the world in all sports, with better training, coaching and diets year after year. So it would be logical to assume that each Games is more competitive than the last. Surely? Perhaps. But when it comes down to actually winning a medal, any medal, might it be getting easier?

Let's look at the number of sports and available medals over time. The Olympics started small: there were just 122 total medals in 10 sports in 1896. There are nearly 1,000 medals on offer now, in 41 sports.

As some sports are fairly new, their pool of contestants will be significantly smaller than other established sports but they all add to the total. In the older sports there are more events to enter. So does that make it easier to win a medal? Not if we assume standards are rising faster, or with some correlation to the overall number of medals available – but that is far from obvious. What is clear is that there are definitely more opportunities to win a medal, whether it is in a new sport, or in a more established sport with a wider range of events. Officially, the number of sports has been capped by the International Olympic Committee. Baseball and softball were abandoned for the London 2012 Games, the first sports to be dropped since polo in 1948. But evaluation and lobbying for new entrants means that the strange evolution of Olympic sports makes winning a medal more of a lottery. Who are the best rock climbers in the world? Or wakeboarders? Both are aiming for inclusion. Of course, it's just as important to know how many athletes are taking part. If the medals available have doubled, but the number of athletes competing has gone up by more, then competition has got harder. However, that's not the case.

Although there are now around twice the number of athletes taking part compared to the Olympics of the 1960s and 1970s, the number of athletes at each Games has levelled off. There were fewer competitors in London 2012 than in each of the previous three Olympics.

The number of athletes per medal started at just over two in 1896 and went up to 11.5 in 1960. Unlike the total sports and medals chart which rises from 1960 onwards, the athletes per medal from that point wavers. Since then, only two Olympics have had more competitors per medal.

The number of athletes at each Games has been affected by the various boycotts in 1976, '80 and '84, and location in '32 and '56 because competitors were unable or unwilling to travel so far. But since 1988 the highest ratio of athletes to medals was actually Atlanta in 1996, with over 12. After that it has trended downwards, with London's 10.9 athletes per medal the lowest since 1984. More medals and fewer athletes per medal might suggest easier competition. What about the pool of talent?

The number of countries taking part has surged. With such a wide talent pool, does that make the Games more competitive? The answer is no. Although there are international Olympic qualifying standards, a large number of athletes are included in the Olympics under 'Universality places'. For instance, the qualifying time for the men's 100m in London 2012 was 10.18s. But the maximum any country can enter is three competitors for the event. Pity then, Mike Rodgers and Darvis Patton, who came fourth and fifth in the 2012 US Olympic time trial with times of 9.94 and 9.96 seconds respectively, good enough to have got them into the London final. However, they missed out on Olympic places - there were 14 runners in the heats who ran a full second slower at the Games; some finished in over 11 seconds.

Arguably, the Games are being diluted by competitors from smaller countries who don't stand a chance. For the men's 100m, there were 85 competitors, 30 of whom were on Universality places. The number of countries taking part is almost at the limit. Pretty much every country in the world that can muster a sports programme and Olympic committee is sending competitors: 204 countries took part in London 2012. There are only 192 country members of the UN.

However, there are still over 70 countries yet to win an Olympic medal, 21 of which have already been to 10 or more Games. Some are relatively large countries: Bolivia, Myanmar and Angola are all still waiting to get someone on the podium. Being rich but small doesn't always help either. San Marino and Monaco are both still medal-less. Despite the new countries taking part, the percentage of countries

More medals to go around

The number of sports and medals at each summer Olympic games

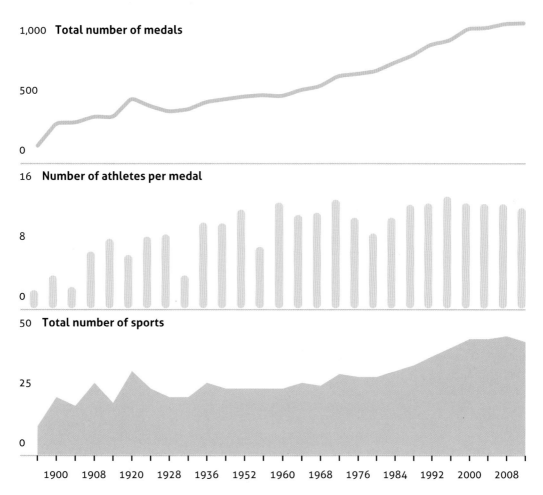

Total number of medals

1,000

500

0

Number of athletes per medal

16

8

0

Total number of sports

50

25

0

1900 1908 1920 1928 1936 1952 1960 1968 1976 1984 1992 2000 2008

not winning a medal has stayed at around 60 per cent. Since 1988, there have been more than 100 countries at every Games that went home empty-handed. The toughest year was Athens 2004: 127 countries won nothing.

Some countries have specialised in a few sports, and have reaped the rewards. South Korea excels at shooting and archery. Ethiopia was the most 'mono-sporting' country at London 2012, with 33 of their 35 competitors taking part in middle- and long-distance running. Kazakhstan was 12th in the London 2012 medal table, largely thanks to its gold medals in weightlifting. These are the examples to follow for countries that want to avoid the fate of Austria, which sent 70 competitors to London 2012 and returned with zero medals,

Portugal in 1992 with 90 competitors and no medals, or the most empty-handed country at a single Games – Spain, with 122 athletes in 1968 and nothing to celebrate.

The gathering together of so many countries to compete on equal terms is an amazing thing, and is a key part of the Olympic ethos. But the generosity ends there. Rather than a larger, more competitive talent pool, there are simply more wild-card have-a-go competitors from smaller countries – some of which are keeping out talented athletes from bigger nations, making the competition harder, not easier.

For all the talk of rising standards, perhaps the competitors in Rio 2016 and Tokyo 2020 might actually have it easier than they think.

Some countries flourish at both Summer and Winter Olympics

Can we learn anything from them?

Which are the truly successful Olympic countries? It's a hard question to answer, because as soon as you start to dig, there are caveats and 'yes-buts' everywhere.

There's historical participation. Take for example, Great Britain. The country is third on the all-time medal table, with 236 golds and 780 medals in total at the Summer Games. Pretty good, you might think. But a large chunk of Britain's medals were won in the infancy of the Olympics, when few countries took part. It also helps that London has hosted the Games three times.

Then there is population to think about: how can Jamaica, with less than 3m people, produce so many great sprinters, when other far bigger countries have none?

Another problem is politics. How do we rank Russia, when many medallists of the Soviet Union would have represented states such as Ukraine or Belarus?

In order to give a more recent picture of success, we have looked at the last five Olympics, both summer AND winter. This is post-Soviet break-up, and a period when many more countries around the world took part. It also allows for the medal boost that comes from hosting the Games to be somewhat diluted.

For each country we have calculated the percentage of medal points won of the total available (three points for gold, two for silver, one for bronze). On the chart we have also shown their relative populations.

What stands out is that the US, China, Russia and Germany are a long way ahead of everyone else when it comes to overall success, with the US the most successful overall, gaining over a fifth of all medal points in both games.

The other countries to score over five per cent of medal points in both summer and winter are France, South Korea, and Italy. (The Netherlands just miss out with 4.6 per cent of summer medal points, and 10 per cent of winter.)

These seven are the most successful all-round Olympic countries of the last 10 games. Others are winter specialists, such as Norway or Austria. Some pick up medals at both summer and winter games, but not enough to be that significant, such as Japan or Poland. Britain and Australia are unsurprisingly summer specialists.

How do countries succeed at both games? Clearly, population helps, along with the other obvious factors such as wealth, facilities, climate and tradition. Doing a few things well is also a good idea. South Korea wouldn't leap to mind as a sporting giant, but it is a very successful all-round Olympic country partly because it has won more medals in archery, taekwondo and short track speed skating than anyone else. It helps to focus.

Hot or cold? Overall Olympic medal success

Percentage of Olympic medal points won (last 5 Games, countries with minimum 5%)

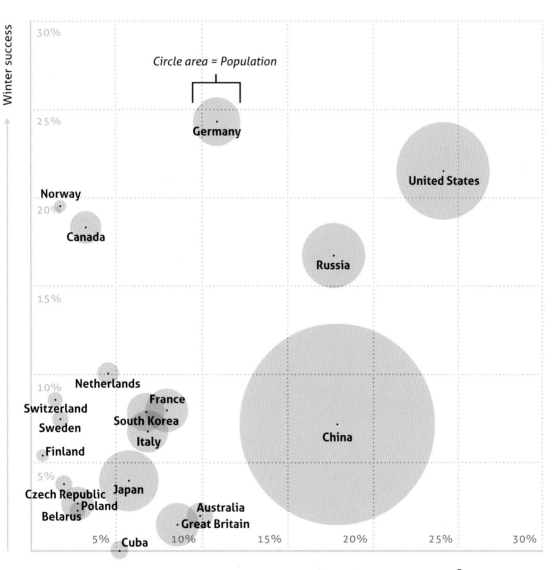

Winter success

30%

Circle area = Population

25%

Germany

United States

Norway
20%

Canada

Russia

15%

Netherlands
10%

Switzerland
Sweden

France

South Korea

Italy

China

Finland

5%

Czech Republic Japan

Poland

Belarus

Australia

Great Britain

5% Cuba 10% 15% 20% 25% 30%

Summer success

Drugs, hosts and obsessions

Why predicting the Winter Olympics is a tough sport

Predicting how many medals each country will get at the Olympics using economic models could almost be an Olympic sport in itself. While the Summer Games attracts plenty of sports economists, few have tackled the Winter Olympics. Perhaps this is because it's a smaller event. Or maybe it's because modelling success at the Winter Games appears to be more dependent on geography and culture.

Whatever the reason, two academics, Madeleine and Wladimir Andreff, had a go at predicting the Sochi Games medal table of 2014*. How did they do?

Not so well, as it turns out. As the chart shows, two countries significantly underperformed – the US and Germany; and two countries did far better than predicted – Russia and the Netherlands. In all, four of the six major predictions were some way off. Only Canada and Norway were within a medal or two of the model. Why?

Andreff and Andreff based their model on previously successful predictions of the summer games, which used GDP per capita and population as two major variables, plus others such as the host effect and country regime. Then they refined the variables to allow for sporting culture, and added in how much snow each country gets, and its winter sport facilities. The disparity between the predictions and results can be explained by two things: the host effect and obsessions.

The host effect is perhaps becoming more pronounced at the Olympics in general. Most studies prior to Beijing in 2008 underestimated the medals that China would win. Similarly, Britain exceeded expectations at London 2012. Then Russia topped the medal table at Sochi in 2014.

However, Russia's performance may not be entirely explained by the host effect. Russia hardly has a clean track record when it comes to doping. In fact, Russian athletics was suspended from international competition in 2015 after the World Anti-Doping Agency (WADA) revealed the widespread and state-sanctioned doping of Russian athletes.

Are Russian Winter Olympians clean? There is no direct accusation of the Sochi medallists. However, Wada said in its report that led to Russia's suspension from international athletics that the lab used in the Winter Games was infiltrated by Russian FSB (the post-KGB Federal Security Service) agents posing as lab engineers, who 'actively imposed an atmosphere of intimidation' on the staff, which compromised the lab's 'impartiality, judgment and integrity'.

Let's just say that the pressures of hosting a mega-sporting event and the desire for home success can have regrettable consequences.

But what of the Netherlands? How on earth did a country with barely a mountain in sight win 24 medals, eight of which were gold? The answer is obsession. It helps to have a sport that the country cares about that isn't mainstream elsewhere. In this case, it's speed skating, an event which requires no mountains, has a long history in the Netherlands, and was backed up with investment that paid off. The country picked up 23 of the 36 speed skating medals available – just under two-thirds, a phenomenal achievement. In previous Winter Games, the Netherlands had won medals solely in speed skating, but simply not to the same extent as at Sochi.

The predicted medals for Germany and the US didn't materialise. Partly this is due to the zero-sum effect: medals picked up by Russia and others are medals the US and Germany can't win. Bad luck may have played a part too. But the high prediction was not a poor call given that they were the top two medal-winning countries in Vancouver in 2010.

With hindsight, perhaps academics are still playing down the host effect. But it's hard to incorporate a national obsession, luck and possible doping into an economic model. As Andreff and Andreff put it: 'predictions must be taken with a pinch of salt'.

---

* 'Economic Prediction of Medal Wins at the 2014 Winter Olympics', Madeleine Andreff and Wladimir Andreff, *Ekonomika a Management*, 2011.

Didn't see that coming

Total and predicted medals at Sochi 2014

 Worse than predicted   Better than predicted

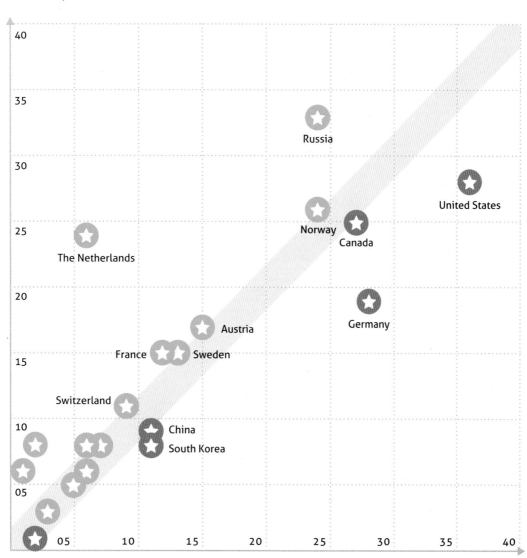

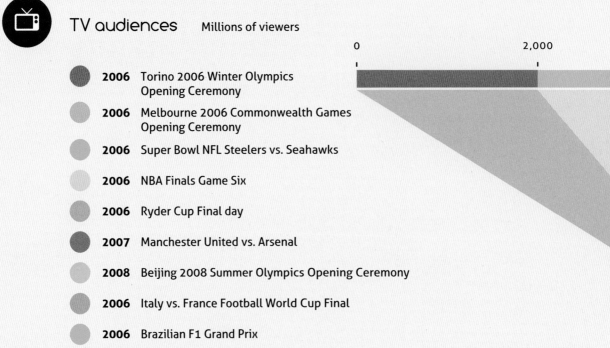

TV audiences Millions of viewers

		0	2,000
2006	Torino 2006 Winter Olympics Opening Ceremony		
2006	Melbourne 2006 Commonwealth Games Opening Ceremony		
2006	Super Bowl NFL Steelers vs. Seahawks		
2006	NBA Finals Game Six		
2006	Ryder Cup Final day		
2007	Manchester United vs. Arsenal		
2008	Beijing 2008 Summer Olympics Opening Ceremony		
2006	Italy vs. France Football World Cup Final		
2006	Brazilian F1 Grand Prix		
2006	Champions League Arsenal vs. Barça		

One billion? Don't believe the hype

Sports administrators are still exaggerating TV audiences.

How many people really watch sporting events? Not as many as we are led to believe, that's for sure. TV broadcasters want to claim as many viewers as possible – or feasible – for advertisers. Sport governing bodies have similar motivations. The real numbers can be shockingly low.

A quick search of news database Factiva shows that an increasing number of sporting events have claimed over 'one billion viewers'. One billion is a nice round number and, as Kevin Alavy of Futures Sport + Entertainment puts it, 'unfathomably high' as the average audience for a single match or race (although not cumulatively for an entire tournament). Futures Sport + Entertainment are an international sport business consultancy who 'use data and analytics to advise the world's most famous clubs, federations and sponsors'.

Futures Sport + Entertainment doesn't reveal its calculations on sporting events. Back in 2006 the *Independent* newspaper published a list of claimed and verified viewers for selected sporting events. The disparity between the numbers given from the two sources is in some cases extraordinary: the opening ceremony of the Melbourne Commonwealth Games had a claimed audience of 1.5bn. The verified audience reported by the *Independent* was just 5m. The chart shows the disparity.

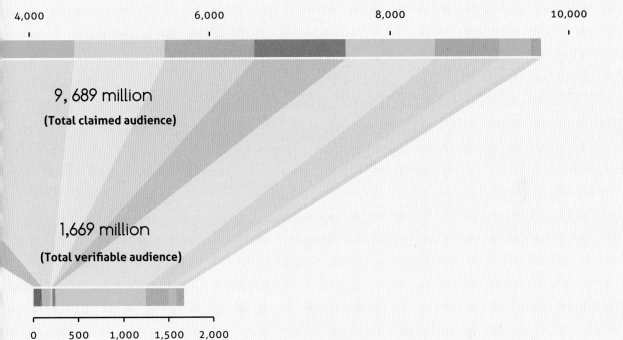

4,000   6,000   8,000   10,000

9, 689 million

(Total claimed audience)

1,669 million

(Total verifiable audience)

0   500   1,000 1,500 2,000

Don't think that opening ceremonies are the most remarkable examples of audience inflation – if anything, they get more viewers than many events because they appeal to both men and women, and as cultural events attract a broader demographic.

In fact, the opening ceremony of the 2008 Beijing Olympics is still the most-watched event in world sport. As Alavy says: 'Nothing has rivalled it. It was a coming-out party, at the right time to get a global audience, broadcast on dozens of channels simultaneously in China.'

So how do TV audiences get measured? There is no set way. Organisations are rarely clear about whether they are talking about viewers for a tournament as a whole (cumulative), or specific matches; whether it's households or total people;

and whether it's for a few minutes or the whole event.

There are no global standards, either. For instance, in the UK, there is Barb, the Broadcasters' Audience Research Board, which is the authority on television audiences. Other countries have similar bodies. But in many parts of the world there's no auditing body, and it's a far harder task.

Alavy suggests that the wide disparity of claimed viewers to actual ones is 'happening less and less, because there is more scrutiny of these figures'. And that would tally with the bumpy chart showing news articles claiming one billion viewers for sporting events. The peak in 2014 was a World Cup year, when Fifa boasted widely that the final would get the magic one billion viewers. We will never know for sure the real figure, but it's probably much lower.

Major League, major problem

Baseball attendance is down, and TV isn't picking up the slack

Baseball has a problem. It is America's game like no other, and as culturally significant as ever, but the fans have just about had enough. More specifically, attendance has peaked, and shows little sign of reviving. At the same time, the costs are going up and up. Something has to give.

The chart shows the average attendance per game for each Major League Baseball (MLB) season since 1916. Since the mid-1970s the trend was for audiences going up year after year. There are bumps along the way as some stadiums were rebuilt and the league was expanded. Overall, the stadium capacity grew alongside the increased average attendance, going up from around 30 to 70 per cent.

Then there were some major stumbling blocks. First, the players' strike of 1994 knocked attendance by a fifth: 25,000 fans attended the average game once the strike ended in 1995, compared to 31,000 before.

The sport took a while to recover, but seemed to be on the up again when the first US recession caused by the Dot-com crash of 2000 dented attendance. But again, gate numbers picked up.

The 2007–08 season, however, may prove to be the high point from which baseball never recovers. Since the financial crisis of 2008, attendance has dropped below 31,000 per game and has shown little sign of increasing.

Although the US economy is out of recession, median household incomes have fallen; in a nutshell baseball fans can't turn up if they don't have the spare money. In contrast, the payroll of baseball teams is going up sharply, as the chart shows, and both the average and median payroll is now over $100m. Who is going to pay for that?

At first glance, you might think it's all TV. Not long ago, MLB teams were given a huge boost by increased TV revenues. Fox and ESPN both signed multi-billion-dollar 8-year deals. Both local and national media deals have seen the proportion of overall MLB revenues from TV go up from 34 per cent to 43 per cent from 2010 to 2014. In the same period, gate receipts and revenues from premium seating have done almost exactly the opposite, falling from 42 per cent to 34 per cent.

The fans are having to dig deeper, too. According to Team Marketing Report, a sports marketing publisher, the average cost of an MLB ticket has gone up from $23 in 2007 to $29 in 2015. The Fan Cost Index score has gone from 177 to 212 over that period. The Fan Cost Index compiled by a Team Marketing Report, that factors in not just the ticket prices for

Fewer fans, richer players

MLB attendance vs. payr

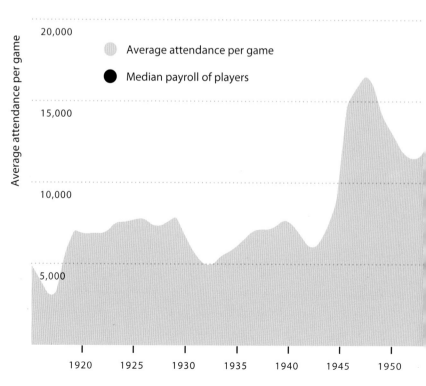

Average attendance per game

20,000

○ Average attendance per game

● Median payroll of players

15,000

10,000

5,000

1920   1925   1930   1935   1940   1945   1950

all teams but also the refreshments, parking, even buying a couple of souvenir caps.

Looking at it another way: from 2010 to 2014 the MLB's yearly payroll went up by over $600m. The fans stumped up an estimated $100m of that at the gate.

So is it all about the fans being at home, rather than in the stadium? TV viewing is declining, too. Viewing figures for games broadcast on Fox and ESPN (the main broadcasters) have slumped, both dropped around a million viewers from 2007 to 2012 – around a quarter to a third of their average viewers. But that's national viewing figures. According to Forbes, there are now 14 out of 24 regional TV markets in the US where local ratings for MLB games now have a larger audience than the nationally televised games. Baseball is becoming more tribal.

Nowhere is this more apparent than in the TV ratings for the season finale, the World Series. Once a game that nearly a third of the country watched, average viewing figures are now regularly below 15 million from a population of over 320m .

Baseball is in a bind. Nationally, TV isn't cutting it in the ratings. Fans are more price-sensitive and increasingly only want to watch their local team on TV. Player wages are spiralling up. At some point, the fun will stop. The next round of TV contracts, due in 2020, should prove interesting.

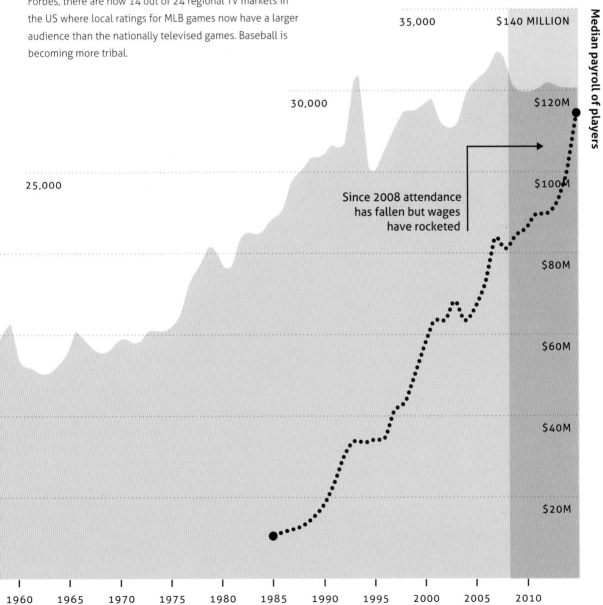

Median payroll of players

35,000

$140 MILLION

30,000

$120M

25,000

$100M

Since 2008 attendance has fallen but wages have rocketed

$80M

$60M

$40M

$20M

1960   1965   1970   1975   1980   1985   1990   1995   2000   2005   2010

Can Moneyball be measured?

Is winning on a budget in baseball so special?

The Moneyball effect ● MLB ● MLB excl. the Oakland Athletics (RH axis) ● Oakland spending (LH axis)

How the Oakland Athletics stopped spending and started winning

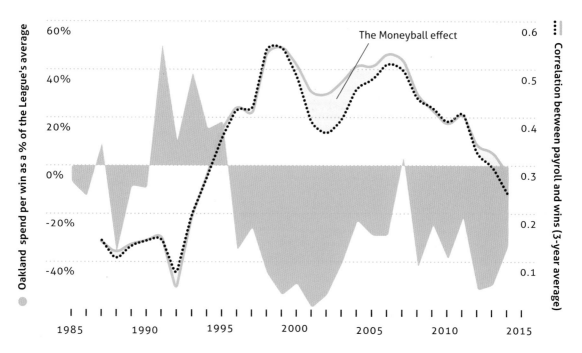

M ention the words 'sport' and 'data' in the same sentence and odds are someone will mention Moneyball inside 30 seconds.

That's because Moneyball is the great sports data story of our time – the tipping point when people started to take spreadsheets and numbers seriously, even if the data contradicted what they saw with their own eyes. Moneyball is also a great narrative, an underdog story, which we all love. But what has the effect been on baseball?

A quick recap: Moneyball is the story of how a baseball team, the Oakland Athletics, used statistics to get an edge on their richer rivals. In sport, including baseball, the bigger spending clubs tend to win. So Billy Beane, the general manager of the Oakland A's, assembled a team of players who had been undervalued for not looking the part, but whose data told a different story.

In 1991, the Oakland Athletics were the biggest spending team in baseball but they didn't even make the playoffs. For the three seasons before, the team had been in the World Series, winning it in 1989. By 1995, the A's had new owners who were not interested in spending big, but being competitive on a limited budget. Beane took over in late 1997 as general manager. The club got to the playoffs in the years 2000 to 2003, and 2006, despite being one of the stingiest teams in terms of player salary.

The green shaded area shows the A's' return on spending each season – it is the team's payroll divided by wins each season, as a percentage of the league average. In other words, it shows how successful the A's' spending was relative to other teams. The lower the spike, the more you can see A's value for money.

In 1991, the A's spent more than 50 per cent more than

other teams to get a win. By 2001, it was the other way around – the A's spent $331,000 per win, compared to the league average of $811,000 – 60 per cent less. And that's compared to the league average – compared to the other playoff teams, the A's were miserly. Three of the other teams were spending over $1m per win. The New York Yankees' total payroll was more than three times higher than the A's'. Overall, the A's were the second-lowest spending team in the league, yet had the second-highest win rate.

How the A's did it is another story. Here, we are interested in showing the effect on other teams.

The waves show how the A's got value for money. What about the rest of the league? Did the other teams catch on?

The black line on the chart shows the correlation for Major League Baseball between all the teams' player salaries (total payroll) and wins-per-game (in other words, percentage of games won per season) for a rolling period of three seasons from 1985 onwards.

It's tempting to construct the following narrative: from 1990 onwards, as TV money pours into the game, the teams that spend big get better results. But then as the Oakland A's prove in 2000 onwards, spending big doesn't always pay off. The correlation rises briefly, then falls as the link between money and results is broken, and everyone wakes up to smell the stats.

Does that hold true? What if we run the same analysis, but without the Oakland A's? How would the league relationship between payroll and results look then? Running the data without the A's – shown by the blue line – we can see that the league correlation for the period of 2000 to 2006 would in fact have been much higher. The A's were responsible for a large chunk of that early break in the link between spending and success in the Major League. In other words, that shaded gap between the lines is indeed the Moneyball effect.

However, other things influence how teams spend. Specifically, in 2003, the league introduced a tax on high payrolls, called the competitive balance tax, usually known as the luxury tax.

Has this had any impact on spending? Obviously, it has potentially given team owners an incentive to keep their spending under the tax threshold. The effect would, however, seem small. Any team that has deep enough pockets can afford to both spend big on players *and* pay the tax. In fact,

the threshold for the tax has been much higher than most teams' payroll. If it were set at a lower rate, you might expect a cluster of teams spending just below the threshold, and then several teams bursting through.

The reality is that just six teams have ever paid the tax: the Detroit Tigers, San Francisco Giants and Anaheim Angels (once each); the LA Dodgers (three times); the Boston Red Sox (seven times); and the New York Yankees (every year). Essentially, the luxury tax is mainly a Yankees tax: the team has paid over a quarter of a billion dollars in tax over the last 12 years.

The correlation between payroll and results went *up* fpr a few seasons, falling from 2006 onwards.

So is the decline in the link between spending and results due to more teams using a statistical approach? Possibly. If the whole league used statistics like the Oakland A's, money would again become the big difference. As economists like to say, *ceteris paribus*: but all other things aren't equal. You can in fact win on a budget *without* being a poster-team for the sabermetric revolution*.

What few people who have read *Moneyball* will realise is that in 2003 there was another team that made the playoffs who spent *less* than the Oakland A's. The Florida Marlins qualified for the playoffs with a 0.562 record, but a payroll of $49m (compared to the A's $50m). The Marlins then went on to *win* the World Series, 4–2 against the richest team of the lot, the Yankees, something Beane's Oakland have not done.

So why isn't *Moneyball* about the Marlins? Because their success wasn't about stats. The Marlins' success was based on a mixture of homegrown talent and a few veterans. Plus, the manager didn't exactly fit the part: he was a 72-year-old called Jack McKeon who didn't believe in sabermetrics, nothing like the young upstart Beane taking on the numerically illiterate dinosaurs.

The Marlins aren't the only example. There was the Minnesota Twins in 2002, another playoff team that spent just $400,000 more than the Oakland A's. The Twins based their team on youth and beat the A's to get to the American League final.

There are other examples. The point is, the Oakland A's aren't the first or only team to win on a budget. But veterans plus youth doesn't have the same ring to it as the scientific triumph of the data nerds.

---

*Sabermetrics was a term coined by baseball writer Bill James back in the 1980s for the study of baseball statistics. The acronym SABR stands for the Society for American Baseball Research.

The winners and losers of baseball's 1994 strike

Overseas players got more of a chance

Baseball is the quintessential American game – but it has never been exclusively played by Americans. The Major League has had players born in over 50 countries, from Afghanistan to Vietnam. And this isn't a recent development: most of the countries on the list had a player represented before the 1960s.

However, despite the list of far-flung countries, the majority of players were from the US. *Sports Geek* has compiled data from *Baseball Almanac* of every MLB player's country of birth and their debut season. The chart shows the percentage of non-US-born MLB players starting out in the league from 1900 onwards.

This doesn't represent success. Some players may come from overseas to play the MLB and be dropped after a single season – others might be in the Hall of Fame, such as Venezuela's Luis Aparicio. Success or career length is irrelevant here. What we want to see is where players are coming from, not what they do next.

You might expect there to be a gradual shift as overseas scouting and better information leads MLB teams to look overseas. And the proportion of non-US born players had been rising, gradually; but there is a big change in 1995. The total number of non-US born players leaps up from 21 to over 50, and (aside from 49 in 1998) never goes below 50 again, and is as high as 72 in 2006. The percentage of US-born players from that point falls below 80 per cent, and has never gone above since. Why? What was special about 1995?

For the answer, we need to look at the year before: 1994. That was the year of the players' strike, which led to over 900 games cancelled, no playoffs, and no World Series. The dispute was complex but boiled down to a rejection by the players of a salary cap proposed by the Major League team owners.

So what happened the following year? Were owners so cross with US players that they decided to look outside the US for players?

Possibly. But for many players, the strike was a watershed

Overseas players: more of a chance?

MLB: Non-US born players (by debut season)

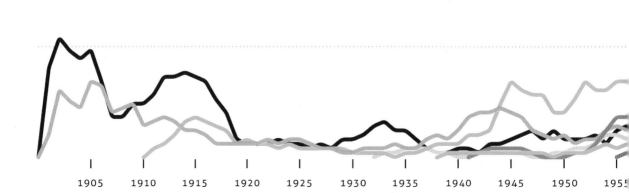

Dominican Republic　　Venezuela　　Puerto Rico　　Canada　　Cuba　　Mexico　　Other

moment, a chance to do something they hadn't done for years – not play baseball. Instead, they spent time with their families and friends. Going back was hard.

As Hall of Fame pitcher Goose Gossage told *USA Today*: 'The strike got me, man. It got a bunch of us. We faded in, and we faded out ... You don't leave the game. The game leaves you.'

Equally, this was an opportunity for others. In the 15 years before the strike, 124 players from the Dominican Republic debuted in MLB. In the 15 years after, 309 played. There was a big increase in Venezuelan players post-strike too.

Of course, overseas players are often cheaper, especially from poor countries like the Dominican Republic. And the strike, which resulted in the players avoiding the proposed salary cap, meant that team owners had an extra incentive to look for bargian players. Now all the MLB teams run academies training young players from the Dominican Republic with the hope that one or two will break through. All MLB clubs are now required to have Spanish translators,

with around a quarter of players from Spanish-speaking countries.

The MLB is nothing like English football's Premier League, with more than half the players from overseas. The 1994 strike was seen as a victory for the players, securing their financial future, but it had other consequences. It ushered in a new international era and crushed many players' careers, and it destroyed the Montreal Expos as a team as well, the best team in baseball that year, but denied a shot at the World Series.

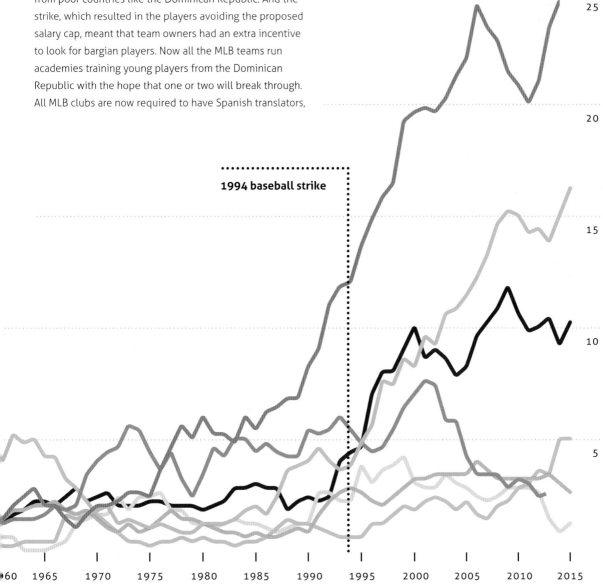

1994 baseball strike

No bases for old men?

Baseball can't make its mind up about speed vs. age

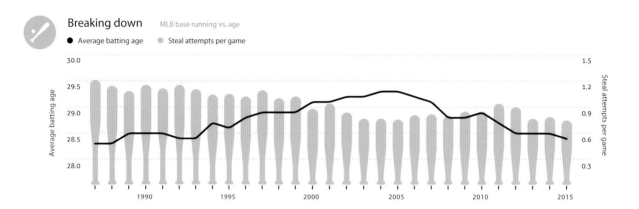

Breaking down   MLB base running vs. age

● Average batting age   ● Steal attempts per game

For a time, it appeared that baseball batters would get older almost forever. From a low of 27.2 years old in 1970, the average age of batters in Major League Baseball rose irregularly and then consistently in the 1990s, until it was over 29, the highest since the Second World War.

While two years average doesn't sound like a lot, it's the equivalent of the entire league not changing any players for two years. Given that every year there are a clutch of new draft picks, there must have been quite a few older players out there. Then the average age started to fall. What drove the increase in age? And what reversed it?

Part of the answer is almost certainly drugs. The 1990s saw a slugfest of hitting, with bulky batters clobbering the ball out of the park time and time again. Steroids didn't just mean big home runs. It also allowed players to stay fitter and stronger for longer in their careers, so older players avoided retirement and the average age stayed high.

Then the Balco scandal erupted, the governing body of Major League Baseball got serious about drugs, and the game changed. In 2002 a Federal government investigation began of the San Francisco Bay Area Laboratory Co-operative. Balco had been supplying steroids and other illegal performance-enhancing substances (including the famous EPO hormone) to top-class athletes for years. Unlike most sports, baseball had no ruling about the use of steroids (and of course no testing) before the scandal.

Drug busts meant fewer big-hitting sluggers. Fewer sluggers meant fewer home runs. And fewer home runs

meant runs (i.e. points) were harder to come by. In order to get runs, it became more important to steal bases – to sneak around the diamond to pick up runs. And that requires speed, not strength. Yet speed tends to decrease with age – hence the trend for younger batters. If we look at the number of base steal attempts per game since 1987, it rises and falls in direct contrast to age. Youth = speed = steals.

Since 2011, however, the batting age and steal attempts have both fallen. Batting age has gone from an average of 28.7 to 28.4. Steal attempts have fallen from 0.94 to 0.74 per game, a low not seen since the 1970s.

Why has the relationship between batting age and steal attempts broken down? Well, coaches are getting cleverer in their use of data. And they have worked out that stealing bases might work now and then, but the risks are high. If you get caught stealing, you are out, and with only three outs allowed per innings, it hurts the team's chances.

How successful are teams at stealing bases? The success rate has risen since 1993 to over 70 per cent, which is seen as a high rate. The peaks of 2007 and 2012, however, of just over 74 per cent, look to have gone. In 2015, it was 70.2 per cent. Stealing has fallen out of favour a little, as it has become less successful. Fielding teams wised up to the fact that stealing was on the increase and became extra vigilant.

Meanwhile, the average age for batters has dropped to the lowest in the last 20 years. It's now a game for young men, even if stealing is less prevalent.

The Balco effect

A blip or a game-changer?

In general, baseball and athletics maintain fairly separate existences. But that all changed in 2001–02 with the Federal investigation into the Bay Area Laboratory Co-Operative (Balco). The stars that were eventually caught were some of the biggest at the time in sport: Barry Bonds, Jason Giambi, Marion Jones, and Tim Montgomery.

When the scandal broke it demolished the notion that widespread organised doping in athletics was just something that Eastern Europeans did, rather than Americans. It also showed how far behind the times baseball was in not having a testing programme at all. Balco was run by Victor Conte, who supplied leading sports stars with various combinations of advanced steroids and other drugs. So what has the effect been on baseball and athletics?

The aim in sprinting is simple: go faster. Steroids help in training recovery as well as helping to build muscle power. In baseball, the main goal is to hit with more power, and to get more home runs. To get an idea of sprint times, let's use the number of sub-10 second 100m times for men, for each year since 1980, on a rolling four-year average.* If we compare that to the number of home runs hit per game in Major League Baseball since 1980, two different patterns emerge. Both see a peak around 2000-01 and then a decline. This

could be from lots of factors: retirements, a collective dip in standards, sub-standard training. Or it could be because the investigators were coming, and athletes and ball players didn't want to get caught. Ask this: how did two completely otherwise unconnected sports see a dip in standards at roughly same time, just as a huge drugs scandal unfolded?

The difference though, is that baseball's drop in home run hitting has continued, while athletics times have picked up again, and gone even further. How come? Firstly, there's Usain Bolt, who has pushed sprint times into a new realm. Athletics is also still blighted by drug cheats, who have served their bans but are back running again. Athletics has shortened drug bans, and athletes can receive reduced bans for cooperating in investigations.

Baseball may or may not still have a wide doping problem, but the era of power sluggers looks over. Pitchers are more talented, defensive systems are better, and teams now have access to data that can be used to nullify a batter. Plus, there's now drug testing, and the MLB has imposed stricter sanctions on those caught. That, and the league has started to favour younger, quicker players.

* A rolling four-year average smooths out seasonal fluctuations, and coincides with the Olympic cycle.

Unnatural peak? The BALCO effect

4-year rolling average

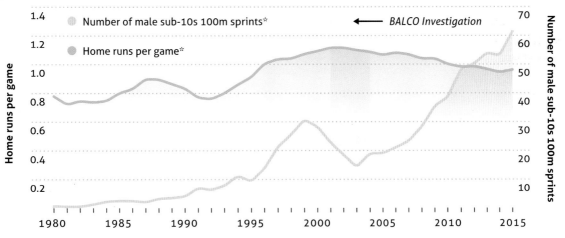

- Number of male sub-10s 100m sprints* ← BALCO Investigation
- Home runs per game*

The end of world records?

Athletics has a progression problem that can't be fixed

There's nothing quite like hearing a commentator shout 'It's a new world record!' So if you hear it, cherish the moment: there are fewer and fewer world records being set.

Athletic progress is slowing down, as the chart shows. It plots a tally of men's and women's world records set in each year since 1945. For men, the peak was in the 1950s. For women, records tumbled in the 1980s. The irregular bumps in the chart are partly due to the Olympics and World Championships which often generate world records.

Since then, the frequency of record setting has declined rapidly. There was only one men's world record set in 2013, the lowest number since 1917. The women have set records later than the men partly because they have started events later, such as the pole vault (1992) and the hammer (1995), which then sparked a flurry of record setting.

Falling fast
World records set each year
● Men ● Women

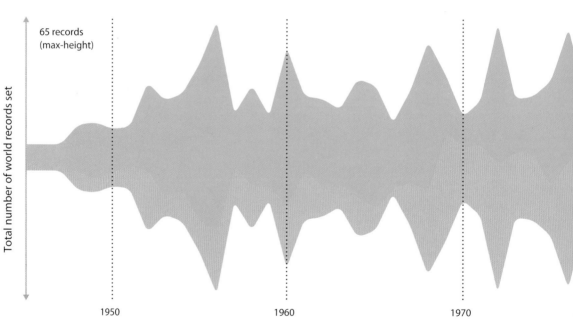

65 records
(max-height)

Total number of world records set

1950   1960   1970

The decline would be even greater if the athletics governing body, the IAAF, didn't keep inventing new disciplines, such as the distance relay or adding previously unofficial events to the world records list, like the half marathon.

Of course, there is a limit to what the human body can achieve; but every so often, the previously-accepted limits are challenged, such as recently in the 100m by Usain Bolt, and in the marathon, as several runners get closer to the 2-hour mark.

It seems though that these shifts in what is possible are infrequent and unpredictable. Will we suddenly see a major improvement in the high jump? It's not impossible, but it is extremely unlikely.

The other problem is drugs. As testing procedures have improved, some records are now out of sight almost permanently. It is no coincidence that 11 of the current women's world records were set in the 1980s, with seven in 1988 alone. As well as the suspicious records in the 100m and 200m held by Florence Griffith Joyner, there are Soviet-era records such as the long jump, discus and 100m hurdles. It's impossible to say that all of these records are drug-assisted. They just haven't been threatened for 30 years. Athletics administrators have openly debated the idea of erasing the record books and starting from scratch.

There is also the question of incentives. You don't get rich doing the long jump. Athletes who might be great jumpers can earn far more sprinting, or using their speed in other more lucrative sports. The men's long jump has only been broken twice in the last 48 years. That's partly because it's technically difficult, and partly because the marks set by

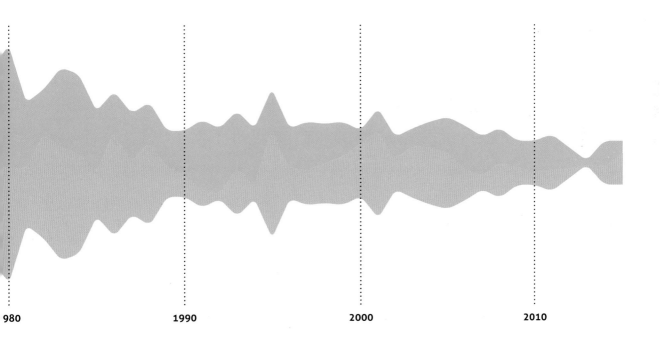

980   **1990**   **2000**   **2010**

Bob Beamon (8.90m) in 1968 and Mike Powell (8.95m) in 1991 are extreme. Could Usain Bolt break 9m? Perhaps, but the risk of injury for little payoff means he's unlikely to try any time soon.

While it is unlikely that world records will dry up completely, they will become rarer. And that is a problem for athletics audiences, who want to see performances at the limit. Which means there are only two options: scrap the old records and start again, or relax the restrictions on drug use. Given the problems athletics has with credibility, neither is going to happen. (Records will continue to be beaten more commonly in long-distance events, however, where there is more room for improvement.)

Count yourself lucky if you see an athletics world record. They are now less frequent than seeing a hole-in-one on golf's main PGA tour.

Why Bolt could do better

The 200m world record is actually a big disappointment

Usain Bolt is the fastest man ever – that's not in doubt. His world records and gold medals in all the major events are proof enough. But he has room to improve when it comes to his 200m world record. Here's why.

Great sprinters can excel at both the 100m and 200m. Some even manage to win the Olympic or World 100–200 double like Jesse Owens (1936) or Valery Borzov (1972). And of course there is Usain Bolt.

Historically, over time, the average speed for the 200m has been higher than for the 100m. The reason seems obvious: 200m is not that tiring a distance for the super-fit, and you have a flying start for the second 100m, whereas the slowest part of the 100m is the first 0–20m. Of course, there is the bend, which slows runners down, but not enough to make the 200m a slower race in metres per second.

Until Bolt came along. Look at the chart of the speeds in metres per second (mps) for the 200m and 100m in the last few decades, and it's pretty clear: over time, the 200m record is faster than the 100m.

One exception is the 2-year period falling between 1994 to 1996, when Leroy Burrell's 100m record of 9.85 put the mps rate higher than the 200m record. Before that, the 100m had just caught up, and was equal to the 200m speed for a few years (to 7 decimal points, anyway). There was also the blip in the chart where Ben Johnson set the 100m world record in Seoul, before being found out as a drugs cheat.

Then Michael Johnson (no relation of course) destroyed the old 200m world record with two quick records in succession, dropping the time to 19.32. That put the 200 mps rate at 10.35, a level that wouldn't be matched until Bolt arrived.

Bolt has set three 100m world records, and two at 200m. But while his last record, the 200m mark of 19.19 seconds, is clearly way ahead of anything achieved by any other runner, compared to his 100m world record of 9.58 it is slow stuff. It's a full 0.01 mps slower. That might not sound a lot, but the chart shows the gap – and historically, the 200m should be faster, not slower. At the rate that Bolt ran his 100m record,

his 200m should be under 19.16s.

So why might this be? Bolt has famously coasted in some races while winning: his previous 100m record of 9.69 was notable in that he clearly slowed to the line and hit his chest in celebration.

Did he coast in the 200m when he set the record in Berlin? Not according to commentators or Bolt himself, although he said at the time he could go faster, maybe sub-19 seconds. That would bring the 200m back into its rightful place as a quicker overall record.

One mitigating factor is that Bolt's record 200m run came just four days after he had set the 100m world record, at the 2009 Berlin World Championships. Tiredness is a fair excuse. But having set five world records in the 100m and 200m in a 15-month period, he has not broken any records since 2009.

Of course, big world records in sprinting are rare – and Bolt's times are exceptional. However, although he is such an outlier (of the 22 fastest 200m ever run, 12 of them are by Bolt), he is clearly capable of going quicker in the 200m.

It might seem odd to call Usain Bolt slow, but when it comes to the 200m, the verdict would be: fair effort, could do better.

For comparison, the 1996 records of Michael Johnson and Donovan Bailey, which were both set in Atlanta Olympics, were compared in 'Who is the Fastest Man in the World?', *Anthology of Statistics in Sports*, 2005.*

The author, Robert Tibshirani, looked at all the different variables, the breakdown of speeds during the two races, the comparison to other competitors in the race and the evolution of the two records. He concluded that a 150m hypothetical race between the two would be too close to call. This is with Johnson's average speed being far higher than Bailey's, at 10.35 meters per second to 10.16 mps.

So what does this mean for Bolt? The 100m-Bolt would thrash the 200m-Bolt at 150m. In other words, his 200m really is slow.

---

*Also appeared in *The American Statistician*, vol 51 no 2, 1997 pp. 106–111

Ground speed

100m vs. 200m records

100m   200m

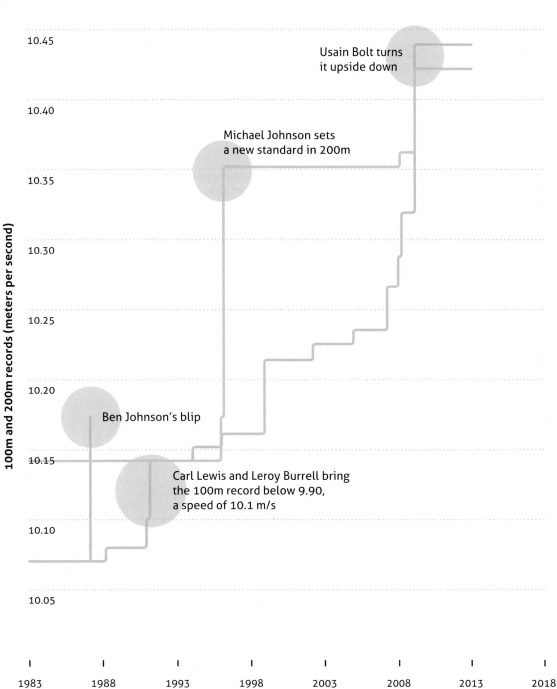

Usain Bolt turns
it upside down

Michael Johnson sets
a new standard in 200m

Ben Johnson's blip

Carl Lewis and Leroy Burrell bring
the 100m record below 9.90,
a speed of 10.1 m/s

100m and 200m records (meters per second)

10.45
10.40
10.35
10.30
10.25
10.20
10.15
10.10
10.05

1983   1988   1993   1998   2003   2008   2013   2018

Flo-Jo's legacy

Why nobody knows who the real female sprint record holders are

While Usain Bolt can be seen as part of the progression of men's sprinting, in the women's equivalent events it's a statistician's nightmare. The 100m and 200m records were set in 1988, and haven't been threatened since. And the spectre of drug-enhanced records means we really can't be sure who is the quickest at all.

The records were set in quick succession by an American sprinter, Florence Griffith Joyner. At the time Americans were known for being great sprinters and not suspected as drug takers. Flo-Jo's rapid improvement from run-of-the-mill performer to extraordinary world-beater did raise a few eyebrows, as did the improvement in her physique. Still, nothing was proven.

She died in 1998 and the dead can't sue, so let's not be shy: Flo-Jo's performances were almost certainly down to performance-boosting drugs. That's the only credible conclusion we can draw from what has happened since.

In a way, Flo-Jo's records were seen at the time in a similar way as Bolt's recent performances – a huge leap forward. But over the years, with no other sprinter coming close, they have become such extreme outliers as to be damning. As fellow US sprinter Gwen Torrence put it in 1995: 'to me, those records don't exist.'

Sprinters tend to get faster over the generations, although usually in ever-smaller increments. Even Ben Johnson's expunged 100m record was overtaken legitimately around a decade later. Flo-Jo's records remain unthreatened a quarter of a century on. For 25 years, the very best female sprinters have trained with ever-improving methods, greater professionalism, more money and greater rewards, but without coming close.

Some of the winning times at the Olympics and World Championships since 1998 makes salutary reading: Veronica Campbell-Brown won the 2011 200m World Championships in 22.22 – nearly a full second slower than Flo-Jo's time.

Sometimes the major events don't always produce the quickest times. A better guide is the top time for the 100m and 200m women's events each year since 1988, regardless of where it is set. The chart shows the best times recorded by a woman for the 100m and 200m from 1989 on, in terms of how many metres behind Flo-Jo they would have been.

The closest anyone has got to Flo-Jo at the 200m is 21.62 in 1998. That was Marion Jones, who was subsequently banned for taking the drug EPO. In other words, the best that the rest can muster in the last 25 years is another drugs cheat who is still more than 2.5 metres behind Flo-Jo. Other years have seen the best 200m runner registering a time that would have seen her come in a full 9 metres back – that was Debbie Ferguson-McKenzie in 2001.

The one ray of hope is the new Dutch sprinter Dafne Schippers, who ran 21.63 to win the 200m at the World Championships in Beijing in 2015. Her performances have been under intense scrutiny, given the history of the event. Still, there is daylight between her and the old Flo-Jo record.

The best 100m runners since Flo-Jo have never got within even a metre of her record. Most years, the best mark has been 2.5 metres behind – and in the 100m event that's a world away.

So who would be the record holder? Next up is Carmelita Jones, who ran 10.64 in 2009, the fourth fastest time ever. Despite being 0.15s slower than Flo-Jo, her time has put her under suspicion. Jones told the *Guardian* in 2010 'I got so much negative press after I ran 10.64 like, "Is she clean? Is she this? Is she that?"... It's unfortunate that I work this hard and I don't get the credit I should get but that's life. There's nothing I can do about it. What do you want me to do? Run slow?'

Jones's quandary is understandable when next on the list is Marion Jones (no relation). Then comes Shelly-Ann Fraser-Pryce – who has also served a drugs ban. And so it goes on.

Such is the paucity of decent 200m times in recent years that two athletes that were the product of a widespread doping programme of the former Soviet Bloc are still close to the top of the list: Marita Koch and Heike Drechsler of the former East Germany. Koch and Drechsler never failed a test, but don't let that fool you: they were never properly tested by today's standards, and documents have shown since that both Koch and Drechsler were given drugs. Drechsler came clean about it, to coin a phrase, in 2001.

Not with the Flo

How the best women's 100m and 200m times compared to Flo-Jo

100m   200m

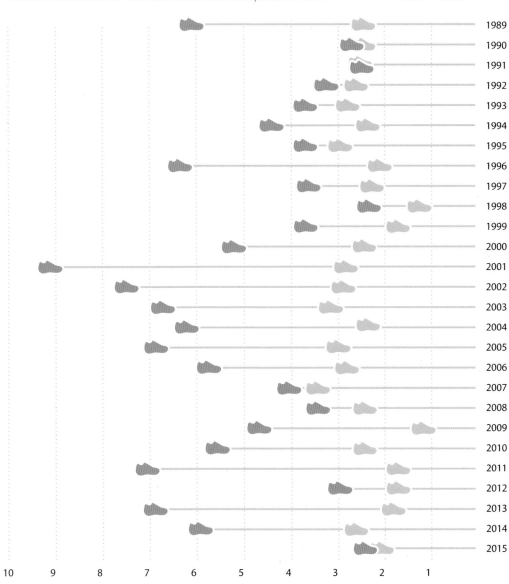

1989	
1990	
1991	
1992	
1993	
1994	
1995	
1996	
1997	
1998	
1999	
2000	
2001	
2002	
2003	
2004	
2005	
2006	
2007	
2008	
2009	
2010	
2011	
2012	
2013	
2014	
2015	

10   9   8   7   6   5   4   3   2   1

Number of metres behind Flo-Jo's record

According to one account, Koch even complained she wasn't being given enough steroids compared to one of her peers owing to favouritism. Koch is still the 400m world record holder – by a long way – with a time set in 1985. The official all-time best 400m list is even more of a drugs roll call than the 100m and 200m, with most of the top 20 quickest times from the 1980s Soviet Bloc.

In fact, only three of the 30 fastest 400m times have been set since 2000. Koch, Flo-Jo, Marion Jones – when it comes to women's sprinting, (clean) progress is hard to find.

Is the 2-hour marathon in reach?

The recent spate of records may be deceiving us

When Dennis Kimetto became the first person to run a marathon in under 2 hours 3 minutes in Berlin in 2014, knocking nearly 30 seconds off the previous record, there was a lot of excitement about whether the two-hour mark could be in sight.

A two-hour marathon would be a huge achievement – a mark that seems as fantastic as the four-minute mile once did. Several commentators have suggested that at the current rate of record setting, we will see the two-hour mark broken some time around 2030.

But we aren't learning from the past. If we extrapolated the marathon records set in the 1960s, we would have expected the two-hour mark to be broken back in 1977. Clearly, extrapolation isn't everything.

So what is behind the current improvements in the marathon world record? As Ross Tucker of the University of the Free State in South Africa points out, marathon runners are no longer 10k runners past their prime – the distance has become more glamorous and lucrative. In fact, the current improvement in the marathon time corresponds with

Is the 2-hour marathon in reach?

Full marathon and half marathon world record times

● Full marathon　● Half marathon

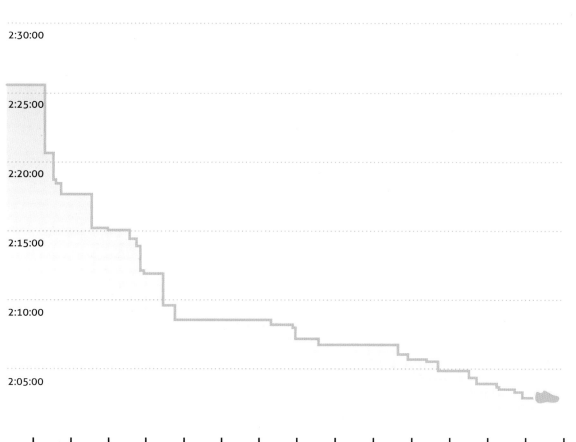

stagnation in the 10k record. The marathon attracts more runners – and more of the best.

With the marathon in the spotlight, and better training methods, you would expect records to fall, but not forever – runners will hit their limit.

Then there is the science. Long distance running comes down to oxygen capacity vs. running economy. No runner has the combination of both at this point to do a sub-two-hour marathon. There is the possibility that we are at the end of a great era for the marathon, and are now about to enter a period of stagnation in terms of improving the world record. We may see a few seconds shaved off every decade or so, but not a continuation of the current pace. That's not how records evolve, normally.

There is another indicator for the marathon: we can look at what the half-marathon tells us.

The half-marathon suggests what marathon runners can achieve (always bearing in mind that runners don't 'scale up' or down perfectly distance-to-distance). Since the mid-1980s, the marathon world record has constantly been just over 2.1 times that of the half-marathon.

The ratio should always be above 2. Assuming that the ratio maintains a level of around 2.1, this suggests that for the two-hour mark to be broken in the marathon, the half-marathon record should be close to 57 minutes. And it's nowhere near that.

A look at the half marathon record progression shows a more consistent pace of improvement. It also shows that to get to 57 minutes, we are looking at many decades, if ever.

This might simply tell us that runners don't take the half marathon as seriously. Or, it might give us a warning sign not to expect the 2-hour marathon for many years to come.

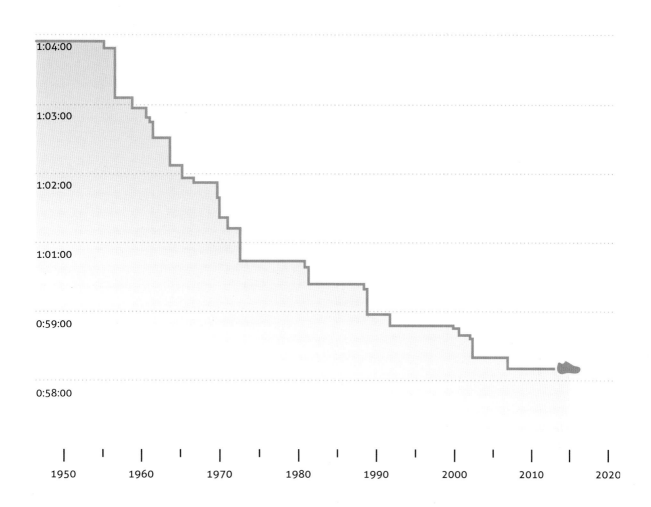

What's small, round and can't be thrown over 70m any more?

The women's discus

Drugs have blighted some athletics events. But none have seen progress halt as dramatically as the women's discus.

The chart shows how discus throws of over 70m have almost totally dried up since 1992. From 1975 to 1992,

there were 161. From 1993 on, there have been only six. Back in 1988 alone there were 43.

How come? Of the 24 women who have thrown over 70m, most are from countries that have what can be best

Women's discus: ruined by drugs?

Throw records by year and country

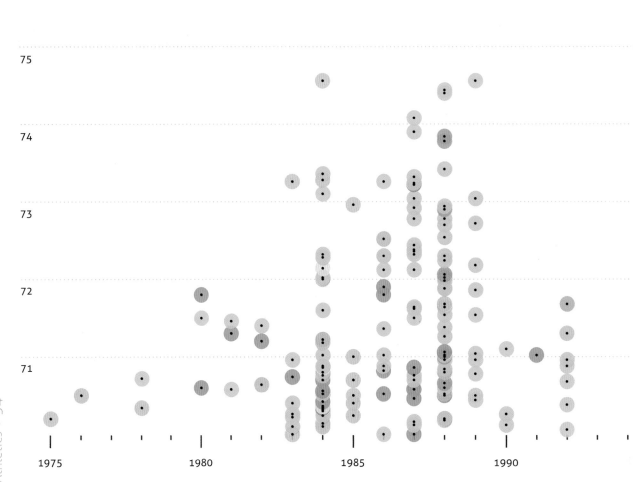

76 metres

75

74

73

72

71

1975   1980   1985   1990

described as a chequered history of drug taking including Russia, Bulgaria and Romania.

In reality, it's all East Germany. The seven throwers on the list from the former Soviet state have thrown the discus 92 times over 70m. Diana Gansky was the most prolific, with 24 throws, although the world record holder is still Gabriele Reinsch with 76.8m, in 1988.

Were drugs involved? Naming no names, the East German athletics programme of systematic doping is well known. It's unlikely that the discus was an exception.

And the latest athletes aren't above suspicion either. Sandra Perkovic of Croatia, who is one of only two women this century to throw over 70m and won the 2013 world title, has served a two-year ban for a banned stimulant. No woman threw over 70m from 1999 to 2014.

If the best throws of the last 15 years can't get close without the use of drugs, what does that say about the 1980s? The only exception is the Cuban Denia Caballero, who threw 70.65m in Bilbao on 20 June 2015.

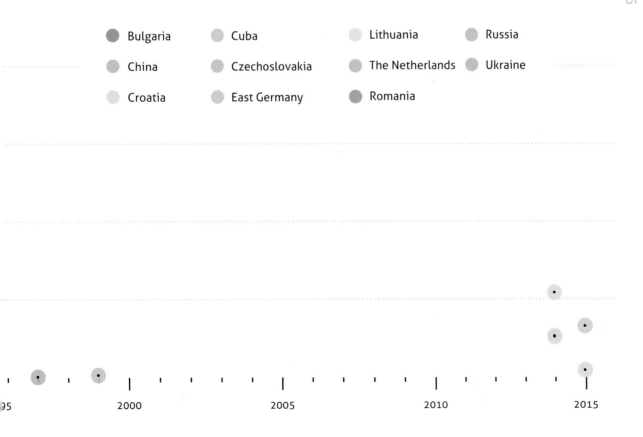

- Bulgaria
- China
- Croatia
- Cuba
- Czechoslovakia
- East Germany
- Lithuania
- The Netherlands
- Romania
- Russia
- Ukraine

95   2000   2005   2010   2015

Switching flags

Which countries are net importers or exporters – and in which sports?

From South Africans playing cricket for England to Bulgarians weightlifting for Qatar, the world is full of players switching national allegiance. It's nothing new: for as long as international sport has been organised, players have changed countries. Some are huge names, such as cricketer Kevin Pietersen. Others are more under the radar. Some move to escape repressive regimes, others for commercial or practical reasons.

Sports Geek is not concerned with the reasons behind players moving: what we are trying to show is where they are coming from, and where they are going, and in which sports.

No comprehensive data exist on flows of players from country to country, so *Sports Geek* has taken a non-official list from Wikipedia, and where possible checked each player movement. However, the data should be taken as indicative, rather than comprehensive.

The chart shows movement by sport, country and number. The criteria for inclusion were a minimum of five players in one sport between two countries.

What the chart doesn't show is cases where a country has seen a big outflow of players in a particular sport but to many different countries, which add up to a big total but don't meet our '5 from 1 to 1' criteria. To counter this, we have added a table simply listing major ins or outs – for example, the number of US basketball migrants, or New Zealand rugby departures.

One thing the numbers show is just how good some countries are at a particular sport, which motivates players to move overseas for international action because they can't get into the national team. It also just shows how talent moves around the world, and where it can find a home.

MAJOR MOVES

OUT	IN
150 Basketball players have left the United States	48 Athletes have moved to France
75 Rugby Union players have left South Africa	42 Rugby Union players have moved to England
50 Rugby Union players have left New Zealand	34 Athletes have moved to the United States
46 Athletes have left Kenya	34 Football players have moved to Equatorial Guinea
37 Baseball players have left the United States	30 Baseball players have moved to the Netherlands
37 Table tennis players have left China	29 Basketball players have moved to the United States
36 Athletes have left the United States	29 Athletes have moved to Bahrain
34 Athletes have left Ethiopia	27 Athletes have moved to Turkey
34 Athletes have left Morocco	27 Athletes have moved to the United Kingdom
26 Athletes have left Russia	27 Athletes have moved to Spain
25 Athletes have left the United Kingdom	27 Rugby Union players have moved to Italy
23 Wrestlers have left Russia	26 Baseball players have moved to Spain
22 Athletes have left Cuba	23 Athletes have moved to Ireland
22 Baseball players have left Curaçao	23 Athletes have moved to Israel
22 Cricket players have left South Africa	22 Baseball players have moved to Italy
	22 Football players have moved to Morocco
	21 Athletes have moved to Qatar
	20 Football players have moved to Palestine

Go with the flow

The movement of players from country to country

Athletics · Basketball · Football · Table Tennis · Weightlifting
Badminton · Cricket · Ice Hockey · Tennis · Wrestling
Baseball · Figure Skating · Rugby Union

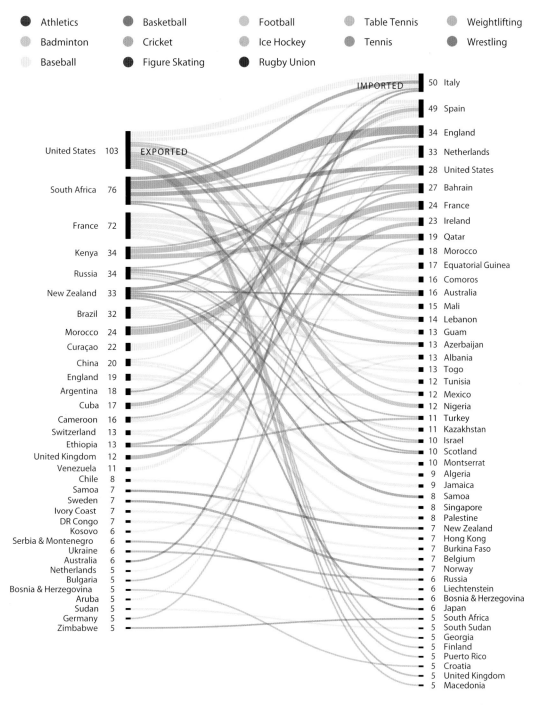

EXPORTED

United States	103
South Africa	76
France	72
Kenya	34
Russia	34
New Zealand	33
Brazil	32
Morocco	24
Curaçao	22
China	20
England	19
Argentina	18
Cuba	17
Cameroon	16
Switzerland	13
Ethiopia	13
United Kingdom	12
Venezuela	11
Chile	8
Samoa	7
Sweden	7
Ivory Coast	7
DR Congo	7
Kosovo	6
Serbia & Montenegro	6
Ukraine	6
Australia	6
Netherlands	5
Bulgaria	5
Bosnia & Herzegovina	5
Aruba	5
Sudan	5
Germany	5
Zimbabwe	5

IMPORTED

50	Italy
49	Spain
34	England
33	Netherlands
28	United States
27	Bahrain
24	France
23	Ireland
19	Qatar
18	Morocco
17	Equatorial Guinea
16	Comoros
16	Australia
15	Mali
14	Lebanon
13	Guam
13	Azerbaijan
13	Albania
13	Togo
12	Tunisia
12	Mexico
12	Nigeria
11	Turkey
11	Kazakhstan
10	Israel
10	Scotland
10	Montserrat
9	Algeria
9	Jamaica
8	Samoa
8	Singapore
8	Palestine
7	New Zealand
7	Hong Kong
7	Burkina Faso
7	Belgium
7	Norway
6	Russia
6	Liechtenstein
6	Bosnia & Herzegovina
6	Japan
5	South Africa
5	South Sudan
5	Georgia
5	Finland
5	Puerto Rico
5	Croatia
5	United Kingdom
5	Macedonia

An anatomy of the greatest try

How 25 seconds of pure magic by the Barbarians unfolded

The 1973 Barbarians try against New Zealand is still regarded as the greatest try of all time. Even after 40 years it still looks fast. It's edge-of-your-seat, brilliant stuff. If you haven't yet, you should watch it on the BBC or YouTube. Other tries have been scored from further back, behind the try line – Philippe Saint-André's try for France against England at Twickenham in 1991 is one. Other tries may have more passes, or be at more dramatic moments – after all, the Barbarians try was at the start of the match, and was in a game that had nothing more than pride at stake.

But there's something about the Barbarians' effort that makes it the best. Maybe it's time, or the teams involved: that sort of thing doesn't happen to the All Blacks very often, and the Barbarians still represent all that's good and romantic about rugby.

It starts with two outrageous sidesteps from Phil Bennett inside his own 22 and ends with a dazzling burst of speed for the line by Gareth Edwards. In between, there are five passes, each one critical and perfectly timed. This, then, is the anatomy of the greatest try.

		Steps	Sidesteps, dummies & tackles broken	Metres travelled	Seconds with ball
A	Phil Bennett	21	5	20	6
B	JPR Williams	5	1 tackle broken	5	2
C	John Pullin	8	1 dummy	12	3
D	John Dawes	18	2 sidesteps / tackles	27	4
E	Tommy David	9	1 tackle broken	16	3
F	Derek Quinnell	5	0	8	2
G	Gareth Edwards	20 + dive for line	2 tackles broken	30	5

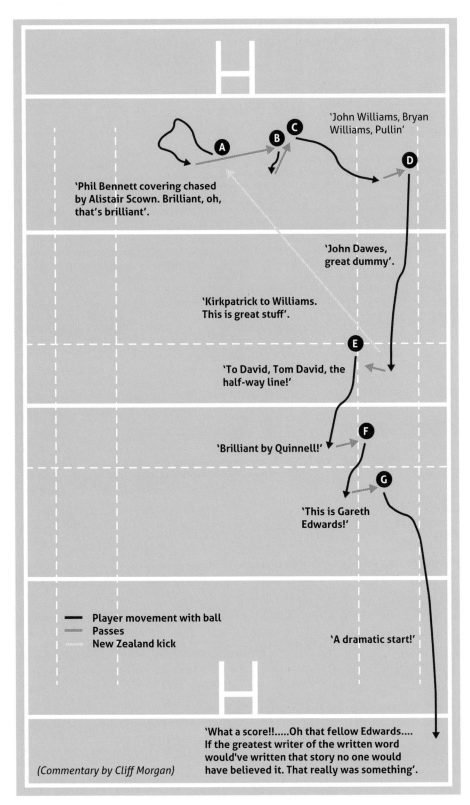

'Phil Bennett covering chased by Alistair Scown. Brilliant, oh, that's brilliant'.

'John Williams, Bryan Williams, Pullin'

'John Dawes, great dummy'.

'Kirkpatrick to Williams. This is great stuff'.

'To David, Tom David, the half-way line!'

'Brilliant by Quinnell!'

'This is Gareth Edwards!'

'A dramatic start!'

— Player movement with ball
— Passes
— New Zealand kick

'What a score!!.....Oh that fellow Edwards.... If the greatest writer of the written word would've written that story no one would have believed it. That really was something'.

(Commentary by Cliff Morgan)

The team

Full back:
JPR Williams (Wales)

Wing:
David Duckham (Eng.)

Centre:
John Dawes (Wales)
Mike Gibson (Ire.)

Wing:
John Bevan (Wales)

Fly half:
Phil Bennett (Wales)

Scrum half:
Gareth Edwards (Wales)

Prop:
Ray McLoughlin (Ire.)

Hooker:
John Pullin (Eng.)

Prop:
Sandy Carmichael (Scot.

Lock:
Willie-John McBride (Ire

Lock:
RM Wilkinson (Eng.)

Flanker:
Tommy David (Wales)

No. 8:
Derek Quinnell (Wales)

Flanker:
Fergus Slattery (Ire.)

Is Southern Hemisphere rugby that much better?

The Southern teams have a better reputation, but the results don't show it

The invincible All-Blacks! The creative Wallabies. The powerful Springboks. Northern Hemisphere rugby has nothing on these three giants of the game.

It's an inferiority complex that has set in for generations, mainly based on some excellent touring sides over the years, and the Southern Hemisphere's domination of the Rugby World Cup.

In terms of results between teams from opposite sides of the globe, it's clear. Mostly the South wins. Not all the time – England's 2003 World Cup win in Australia and the matches leading up to that victory showed that it's not all one-way traffic. But on the whole, South beats North.

The disparity is not just about who wins. For some, it's an article of faith that the Southern Hemisphere sides play more attractive rugby: lots of tries, running at pace, better passing and an overall better spectacle. There's none of the turgid rubbish you see between the European countries on a wet February afternoon.

Is this true? If we look at the history of the Tri-Nations, now renamed The Rugby Championship, the Southern Hemisphere equivalent of the 5/6 Nations, and compare the two events over the same period (1996 onwards), what can we tell from the numbers?

The underlying assumption is that tries are good entertainment, and penalties are dull. Who wants to watch a match decided by kicking the ball over the posts? Of course, a last-minute penalty kick to decide a match is extremely exciting, but in general, a match featuring lots of penalties and few tries is not such fun to watch.

So we should expect the South to have more tries and fewer penalties per match.

Yet over the 19 years of the Tri-Nations and Rugby Championship there have been an average 4.4 tries per match – which is only marginally higher than the average 4.1 per match of the 5/6 Nations. To put that in perspective, you'd have to watch three extra matches just to see one try more. That doesn't add up to a significantly better spectacle.

So what about penalties? The South have scored an average 5.6 penalties per match, slightly higher than the North's 5.3. That, if anything, hurts the South's claim for better entertainment. And there's another factor too: the South have a worse rate of conversions – the 2-point kick at goal after scoring a try – than the North, suggesting Southern Hemisphere kickers are less accurate. They convert tries at a rate of 69 per cent, compared to the North's 74 per cent. In other words, you're more likely to have more penalties scored *and* more penalties missed in the Southern Hemisphere.

Combined, the comparison of tries and penalties tells us that the difference is tiny – certainly nothing to back up the 'better rugby' claim for the South.

But maybe Southern Hemisphere rugby is better to watch because the teams are more closely matched? Do you get more interesting games? Let's take a look.

Overall, the Southern tournament features slightly closer games, with an average points difference per game of 12.4 to the North's 15.2. The average winning scores are very similar: in the South, the winning team averages 29.0 points to the North's 29.5; the losers average 16.6 in the South, 14.3 in the North.

Average scorelines are one area where the Southern Hemisphere looks slightly better value for spectators in terms of competitiveness. But how predictable are the matches? In the North, the home team wins 59 per cent of the time (since 1996). In the South, it is 67 per cent. In one Tri-Nations season (2004) all the matches were won by the home team.

Why should that be, if the teams are closer in average points? One reason might be geography. The Southern Hemisphere teams have to fly tens of thousands of miles to play each other. The furthest teams in the 6 Nations are Scotland and Italy, just over 1000 miles apart. Long distance travel is tiring, however well you prepare.

Is Southern Hemisphere rugby a far better spectator experience? The results simply don't add up.

Northern vs. Southern Hemisphere

Scoring comparisons between major tournaments

5/6 Nations
Northern Hemisphere

Tri-Nations / Rugby Championship
Southern Hemisphere

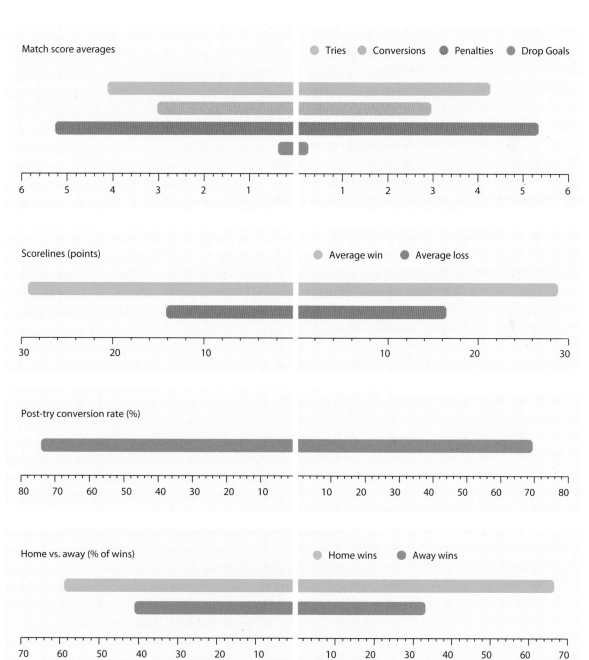

Match score averages

Tries   Conversions   Penalties   Drop Goals

6   5   4   3   2   1   1   2   3   4   5   6

Scorelines (points)

Average win   Average loss

30   20   10   10   20   30

Post-try conversion rate (%)

80   70   60   50   40   30   20   10   10   20   30   40   50   60   70   80

Home vs. away (% of wins)

Home wins   Away wins

70   60   50   40   30   20   10   10   20   30   40   50   60   70

How important is home advantage?

Rugby results in the 5/6 Nations have changed less than you think

International rugby is far more unpredictable these days, isn't it? Home advantage counts for far less as teams have become more professional, better organised, and players are more familiar with travel. Or so the theory goes.

Let's look at a simple metric: home wins. The longest running rugby tournament, the 5 Nations, now 6 Nations, since 1947 is our dataset.

What do we see from looking at the home win rate per year? Quite a lot of variation, in fact.

The chart shows that the two periods when home advantage counted least were in the mid-1960s and late 1990s. Playing at home was most beneficial in the mid- to late 1970s. The only season when every home team won was 1973.

How important is home advantage?

Percentage of 5/6 Nations home wins per year

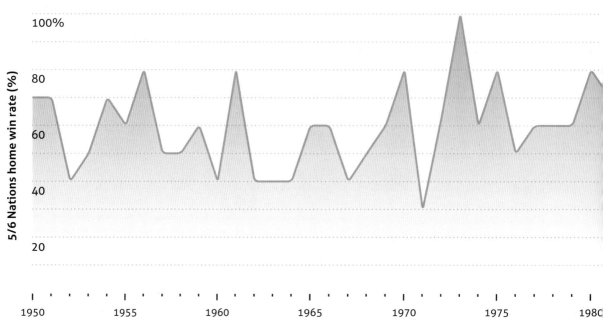

Even the admission of Italy in 2000 into the tournament didn't change the home advantage rate that much, despite Italy losing a lot at the start. Italy have now won 10 of their home matches in the 6 Nations, not that far behind Scotland on 14.

Overall, the average and most frequent or modal number of home victories per season is 60 per cent, or 6 out of 10 matches in the 5 Nations, and 9 out of 15 in the 6 Nations. When someone says home advantage doesn't count for much these days, the reply should be: 'It never really did.'

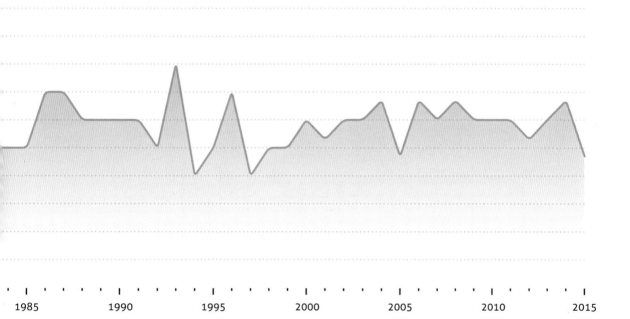

| 1985 | 1990 | 1995 | 2000 | 2005 | 2010 | 2015 |

How the killer penalty takers changed rugby

Has the accuracy of goal kickers kept players honest?

Rugby is a game of instinct, and great physical demands. But in the heat of battle, there are also calculated risks. Players know how a referee will apply the laws, and adapt accordingly. Sometimes, when a team is close to scoring, an opposing player may commit a foul, deliberately killing the ball, and concede a penalty. Conceding a kick for three points is better than a possibly seven. But if penalty takers become more accurate, the risk and reward from giving away penalties starts to change.

There is no reliable data on missed penalties from rugby archives, only a list of penalty kicks scored. So what can we use instead of penalty accuracy? Well, we can use the number of conversions scored following tries. This isn't a perfect substitute measure: the conversion is taken in line with where the try was scored, and attackers will always look to get as close to the posts as possible. But it's not a bad

proxy – penalties are awarded wherever the foul took place, and many tries are scored in the corners as well as under the posts.

The data is taken from every match since 1947 between the 10 major teams of world rugby: England, France, Ireland, Italy, Scotland and Wales; Argentina, Australia, New Zealand and South Africa. The matches include everything from the two yearly international tournaments, the 6 Nations and Rugby Championship, as well as the World Cup, and other one-off internationals. In all, that is over 1,800 matches.

50 years ago, kicking accuracy was around 50 per cent; now it's nearer 80 per cent. That kind of kicking accuracy (albeit in this example from conversions) makes a big difference to the defending team in terms of judging the risk of giving away penalties. So how has that affected penalties?

Looking at the absolute number is not as useful as it may

Kicking a little honesty into the game

Accuracy and importance of penalty kicks

● 5-year conversion rate (%)   ● Importance of penalties (% of points via penalty kick)

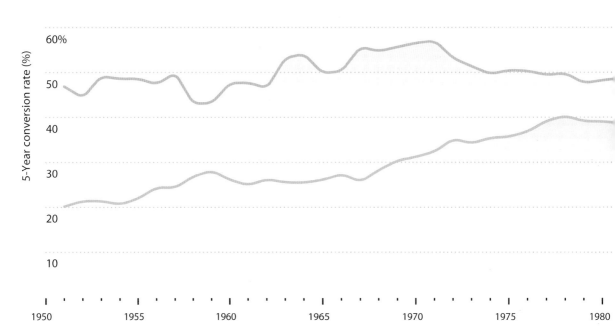

seem: it doesn't tell us about how important penalties were in the context of a match. Plus, as various changes are made to the rules and scoring system, some years you may get more penalties than others.

Instead, it is more revealing to look at the proportion of points in a match that come from penalties. If, for example, four penalties are scored alongside ten tries, that's far less important than a match with four penalties and two tries. Then, to smooth out the yearly fluctuations, we have looked at the five-yearly rolling totals.

What do we see? The chart shows how kicking accuracy stayed around the 50 per cent mark from the resumption of international rugby post-war to the mid-1980s. Then, it went up in two spurts to over 70 per cent in the early 2000s, and it has climbed steadily since. The penalty points contribution climbed from 20 per cent to over 40 per cent in the mid 1980s, probably because kickers became more accurate. But then it drops off from the mid- to late-1980s. There is again a big divergence from 1995 onwards, where penalties drop in importance, and kicking gets even more accurate.

Higher accuracy should mean more points from penalties, if the number conceded remained the same. But instead, more tries are being scored as a proportion of points. Why?

One major change was in 1992–93, when a try became five points instead of four. This clearly pushes down the points accumulated by penalties. Or, you could argue that teams might concede more penalties, as they would be more willing to sacrifice three points instead of a possible seven.

There are other factors: rugby is constantly changing the rules, which often has unintended consequences. Line outs, scrums, rucks, mauls have all had new rules in the last ten years, which has changed how teams play. There are so many variables; it is impossible to say with any certainty. The use of the sin-bin for some deliberate penalties has also changed the risk factor of impeding the opposition scoring.

As it stands, the contribution that penalties make to rugby scores has remained steadily in the 30 to 40 per cent range since the peak of the mid-80s. Meanwhile, kicking gets closer and closer to 80 per cent accuracy. Good kickers are keeping rugby players honest – relatively speaking.

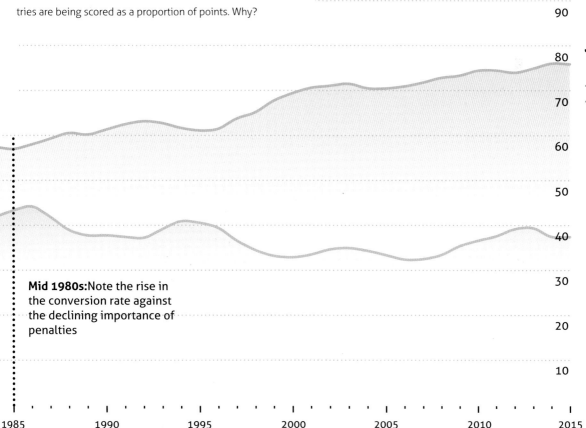

Mid 1980s: Note the rise in the conversion rate against the declining importance of penalties

When the Lions play, who wins?

Answer: France

It's often said, but there's nothing quite like a Lions tour. The four-yearly cycle gives them a similar cachet to a World Cup or Olympics; there's the unique circumstances of the fierce rivalries of England, Ireland, Scotland and Wales set aside to create one squad; and the additional challenge of a southern hemisphere giant taken on in their own backyard. The Lions is a great story.

But Lions tours don't always work out well. Since 1968, when tours have focused on a single country, they have won just 5 of 13 test series. Tours have been marred by violence and gamesmanship, or have been one-sided and poorly organised.

One of the most contentious areas is the country split of the squad. The five tours from 1989 to 2005 saw a shift towards England supplying a large proportion of the players, culminating in the large and unbalanced 2005 squad, where English coach Clive Woodward took 21 fellow countrymen to New Zealand, despite Wales winning the 6 Nations Grand Slam (a clean sweep of wins) earlier that year. The plan backfired, with an embarrassing 3–0 test thrashing. Since then, Ireland and Wales have provided the largest group of players.

Despite those worries of English dominance, it's not clear that an unbalanced squad is necessarily a problem. The 1993 and 1997 tours both had over half the squad from England in the original selection (excluding replacements), and had opposite test series results. Wales provided over 40 per cent of players in 2013, which was also a successful tour. 2005 aside, the composition of the squad broadly reflects the balance of rugby power in the 6 Nations – Scotland is providing fewer and fewer players over the years, reflecting the country's recent record.

Overall though, the team that really benefits is: France. The French players all get a summer off while the best of the home nations get smashed to pieces in Otago, Brisbane or Transvaal. When the 6 Nations comes around the following year, French players are fresher – and it shows.

It's very clear in recent years: 1998, 2002, 2006 and 2010 saw France victories (including three Grand Slams) in the 6 Nations and they were in contention until the last match of the 2014 event. Given the tougher physical demands of the modern game, it looks like advantage France.

In the 5 and 6 Nations Championship since the Second World War, France have won a total of nine Grand Slams, four of them post-Lions tour years. If that sounds fairly even, it's not: there are of course far more years when the Lions don't tour. It actually works out as 5 Grand Slams in the 46 years that don't follow a Lions tour, compared to 4 Slams in the 18 that do, which works out as 11 per cent and 22 per cent. Effectively, France has won twice as many Grand Slams in post-Lions tour years compared to other years.

But it's not just about Grand Slams – the 5/6 Nations Championship can be won without a clean sweep of victories. Here it gets complicated. There are outright tournament wins (which include Grand Slams), and shared wins. If you look at the years that didn't follow a Lions tour, France won the championship outright 22 per cent of the time, with 10 out of 46. In post-Lions tour years, it's 7 out of 18, or 39 per cent of the events. That's a big difference.

The shared championship wins were awarded in the 5 Nations before teams were separated on the difference in total points scored and total points conceded. If we go back and look at who would have won under those circumstances, France get an extra four wins – meaning their championship total would rise to 14 in 46 years, a rate of 30 per cent – still clearly behind their win rate in post-Lions years.

However you cut it, France do better after a Lions tour. The money should be on France winning the 6 Nations in 2018. The Lions are touring New Zealand in 2017 – and haven't won a test series there since 1971.

Lion tours: who goes, who wins?

England   Ireland   ● Scotland   Wales   **GS** = Grand slam

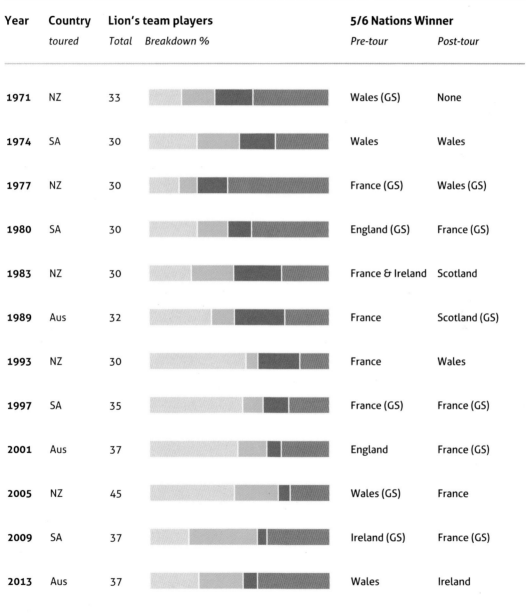

Year	Country toured	Lion's team players Total	Breakdown %	5/6 Nations Winner Pre-tour	Post-tour
1971	NZ	33		Wales (GS)	None
1974	SA	30		Wales	Wales
1977	NZ	30		France (GS)	Wales (GS)
1980	SA	30		England (GS)	France (GS)
1983	NZ	30		France & Ireland	Scotland
1989	Aus	32		France	Scotland (GS)
1993	NZ	30		France	Wales
1997	SA	35		France (GS)	France (GS)
2001	Aus	37		England	France (GS)
2005	NZ	45		Wales (GS)	France
2009	SA	37		Ireland (GS)	France (GS)
2013	Aus	37		Wales	Ireland

0%   20%   40%   60%   80%   100%

Are rugby players bigger than ever?

The major changes in player physique happened
a long time ago

Getting more powerful

Rugby positions by BMI

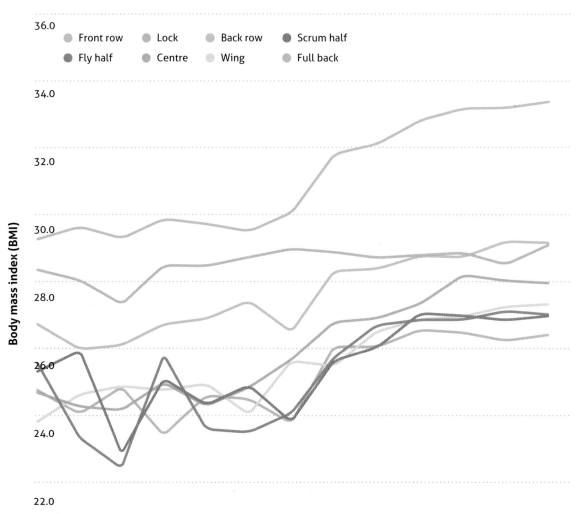

Every so often a rugby union player will take the pitch that seems bigger and scarier than those who have come before.

Sometimes they are toweringly tall, such as England's second row, Martin Bayfield at 6'10" (208cm). They might be huge 20st props, like South Africa's Os du Randt (127kg). They might even redefine what size certain positions can be, such as Jonah Lomu, the New Zealand wing who was 6'5" and 18st 10lb (196cm, 119kg).

But these are the exciting outliers. What we want to know is how players in general are changing size. It seems pretty obvious that rugby players have got bigger, just from looking at footage of previous generations. The question is whether players are 'bigger than ever', as we are still told. Is our perception of a new breed of giant players in line with reality?

Sports Geek has created a dataset of every international rugby player since 1900 to play for the top 10 international teams (Argentina, Australia, England, France, Ireland, Italy, New Zealand, Scotland, South Africa and Wales). The data is from ESPN's publicly available records for each player, which lists their position, weight and height.

We have grouped the data into three positions for the forwards: front row (props and hookers), locks (or second row) and back row (flankers and number 8s). The backs are grouped as scrum-halves, fly-halves, centres, wings and fullbacks. For each position, we have taken the height and weight for players who made their international debut in a five-year period (1970 to 1974, 1975 to 1979 and so on). So what are the main findings?

Forget the talk of massive modern players. The biggest change in height and weight across all positions was 30 years ago, in the 1980s. (There were changes in decades before that, but the data from pre-1960 is too small in sample size to get an accurate picture.) The new rugby players of that decade jumped in size compared to the 1970s, from the front of the forwards to the back of the backs. A typical front row player had a whole stone of extra bulk. Fullbacks were 13 per cent heavier.

The strange thing is that rugby union didn't become a professional sport until 1995, when most of the big size gains had already happened. Professionalism didn't usher in a new era of big player as much as it confirmed what already existed. For instance, second rows – the biggest players on the pitch – have barely changed in both weight and height since the second half of the 1980s.

That's not to say that current players aren't bigger. They are, but the changes are more incremental now.

As players have grown, some extraordinary comparisons can be found. Back row players of the 1990s were as big as second rows of the 1960s. Typical centres from 2005–09 were as big as back row forwards of the late 1970s. And by the late 2000s, props weighed the same (17st 10lb, 112kg) as second rows, despite being on average six inches shorter.

Which brings us to the next change: body mass. Body Mass Index is not always helpful when applied to the general population, as it doesn't distinguish between muscle and fat. This leads to the paradox that extremely fit people can be in the overweight or even obese category. However, for our purposes, BMI is actually quite useful – as a proxy for power.

As players get taller, they are usually heavier. When a typical player goes from being 6' to 6'2", and from 180 to 190lb, their BMI is unchanged. Are bigger players necessarily more powerful? If their BMI stays constant, they may be a bigger presence, but their power has not significantly changed. And rugby players need power to do things like break tackles and to push the opposition back in the scrum.

We can assume that any gain in rugby player BMI is muscle, not fat. So which positions have seen the greatest increase in BMI?

The biggest change is in the front row – the props and hookers. These players have become taller, but their weight has increased significantly more. Front row BMI has gone from just under 30 in the 1970s to over 33 today. It is the only positional area where BMI since 1970 has always risen.

Other rugby positions have seen leaps in BMI as well, mainly in the 1980s. Interestingly, as BMI has gone up, BMI scores have converged around 27 to 29. It is the front row that has split away, far ahead of other positions. In other words, rugby players look more and more alike in terms of build, except for props and hookers.

The changes in rugby physique are real: players are bigger than ever. But the biggest shift happened 30 years ago. And don't be fooled by the occasional giant that comes along on the wing or in the centre. The major change in power and size is in the front row.

The battle for third

The football World Cup and the Olympics are without question the biggest sporting events in the world. What's next?

Consider the following. The Ryder Cup. The World Athletics Championship. The Rugby World Cup. The Cricket World Cup. The Asian Games. The Maccabiah Games.* The Commonwealth Games. The Tour de France.

Nobody doubts that the World Cup and the Olympics are the biggest events in world sport. But the number of events claiming to be (or described in the media as) the next biggest is impressively long. They can't *all* be third.

The problem is defining what we mean by 'bigger'. Audience? Revenues? By countries or competitors taking part?

It's clear from the contenders listed so far, we aren't comparing like with like. Some, like the Tour de France, are yearly. Others take place every two or four years. That makes sensible comparisons almost impossible.

So let's narrow the field with a few criteria. Let's make the decision (you might disagree) that to be the third biggest sporting event after the World Cup and Olympics, it must

also be held on a four-yearly cycle. That excludes the Ryder Cup, World Athletics Championship, and yearly events such as the Tour De France and Wimbledon. This also excludes matches that are the culmination of a whole season, such as the Super Bowl or Champions League final.

Next, we need something to measure. Here's where it gets trickier.

Let's start with the audience. Overall ticket sales is one guide, but that is limited by stadium capacity and the number of matches, which are determined by the host country or city and the format of the event.

What about TV audiences? This area is even more of a mess. Any number put out by a sporting body should come with a handful of salt.

Several events have claimed to have an audience of 'billions'. But is that peak viewers, potential reach, cumulative audience, or something else? There is no standard measure, and many of the figures put out in press

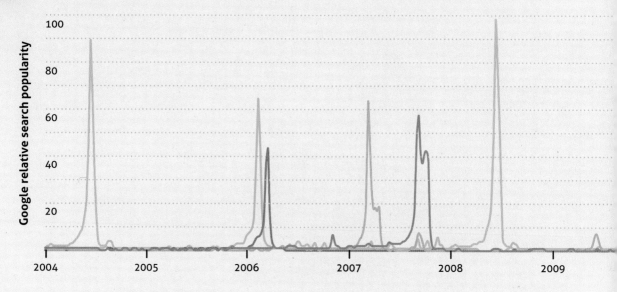

The battle for third   ● UEFA European Championship   ● Rugby World Cup   ● Cricket World C

Total searches for sports events relative to the total number of Google searches over time

Google relative search popularity

100
80
60
40
20

2004   2005   2006   2007   2008   2009

releases are pure guesswork. Kevin Alavy of Futures Sport + Entertainment, a company which specialises in measuring sporting audiences, told *Sports Geek*: 'There's a heck of a lot of exaggeration going on.'

Alavy also pointed out the problem with any audience figure based on 'potential reach': 'If I stream on YouTube a table tennis match between me and my wife, it has a potential audience of billions – anyone worldwide with web access. It's meaningless.'

So if we can't trust the audience figures as a measure of interest, what can we trust? It would be great to find some tool that accurately shows how interested people are over time, a way of capturing what people are thinking about when they want to discover more.

Luckily, a large proportion of the world's population voluntarily participate in such a thing. It's called Google. And Google allows us (via Google Trends) to view its historical database to show how relatively popular different search terms were.

While this is far from perfect in terms of country weighting, language and so on, it at least gives us an idea of how some events stack up.

The chart compares the search topics (which are aggregates of relevant search terms) for the Uefa European Championship, the Winter Olympic Games, Cricket World Cup, Rugby World Cup and Commonwealth Games. Other four-yearly events are not shown as their interest barely registered compared to these events.

There is a clear winner: and it's the football. While the Euros (as it is usually known) is not a world event, and plays second fiddle to the World Cup, the event outstrips all the others by some margin. Euro 2008 and 2004 are both more popular on Google than any edition of the other four events. The Rugby World Cup 2015 is next, just ahead of the Winter Olympic Games of 2010 and 2006.

The Cricket and Rugby World Cups might have greater claim in terms of world population represented (especially India in the cricket), but the Euros has more people searching, by some margin. Football wins again.

---

*The Maccabiah Games is an international Jewish multi-sport event held every four years in Israel, involving 9,000 athletes from 78 countries. It is described on Wikipedia as 'the third largest sporting event in the world'.

● **Commonwealth Games**   ● **Winter Olympic Games**

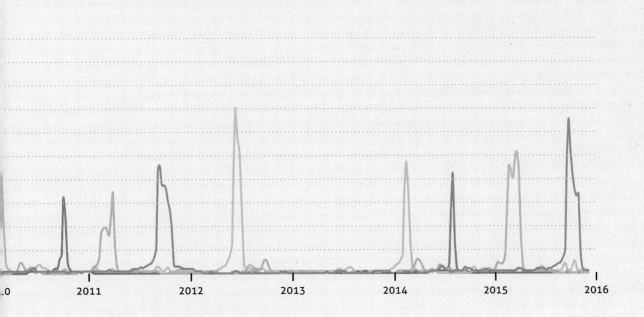

The rise of the quarterback

Has the NFL become too pass-orientated?

When a young Dan Marino, in only his second NFL season, threw for a total 5,084 yards in 1984, it was an astonishing feat. Only five quarterbacks had ever had a 4,000-plus yards season before. To reach 5,000 yards was another level altogether.

Marino's record stood for 27 years. Drew Brees of the New Orleans Saints came close in 2008, just 15 yards behind. But then there was a surge in the 2011 season: Brees and Tom Brady of the New England Patriots flew past the record, passing for 5,476 and 5,235 yards respectively. Peyton Manning of the Denver Broncos pipped Brees' record by one yard in 2013.

A look at the chart counting 4,000-plus-yards seasons for NFL quarterbacks tells the story. In the 1990s, quarterbacks threw for 4,000 or more yards 22 times. In 2015 alone, there were 12 – that's over a third of the league's starting quarterbacks.

We are in the age of the great quarterback – and a new era of passing. Running is dead. Or is it?

We are clearly in an age with great quarterbacks in absolute terms: of the 142 times a quarterback has registered over 4,000 yards in season, Brees, Brady and Manning count for 32 on their own.

Compare that to the earlier era of quarterback greats: John Elway got over 4,000 yards just once for the Broncos. Joe Montana, Hall of Fame quarterback for the San Francisco 49ers, never did. Something has changed.

To pass or run is the basic question on any play in American football. And the trend is clearly towards passing the ball more than running. If we look at the number of pass attempts per game since the AFL-NFL merger of 1966, it has risen from a low in the 1970s of around 24 attempts per team per games, to around 35 now.

This is overstating the change to passing: if we look at passing and running as a percentage of attempts, there was a big shift in the 1970s, but since 1995, the balance has been stuck at around 55 to 60 per cent passing of all offensive plays. To say this is an era of passing and nothing else is wrong: that's to ignore over 40 per cent of the game for the past two decades.

Passing is clearly more effective at moving the football closer to the endzone: you get more yards per play on average. And the average yards made passing per game has gone up a lot. This stat has got many commentators worked up. The *Boston Globe* said in 2011 that the 'passing game is spiralling out of control'. In the same year, the *Bleacher Report* said passing had 'taken over the League'.

The argument was that new rules protecting receivers and quarterbacks had made passing easier. The quarterback's legs are better protected by the rules following an injury to Tom Brady in 2008, for example. Defenders were called for illegal contact on receivers 3.28 times per game on average in 2015, in comparison with 1.19 in 2013. Additionally, running backs were coming up against bigger and fitter defensive linemen, which made running the ball harder.

Looking at passing in isolation is useless. Instead, consider the percentage of yards made per game made from passing (vs. running). It has gone up, as you would expect but not by that much, and it has been fairly static since 1994 – again, two decades of little change.

So is passing in vogue as an idea, rather than a reality? Those big 5,000-plus-yards seasons made 2011 look like a new era. But the passing/running charts show that it wasn't. They were outliers.

One other factor is that quarterbacks may be passing more, but they are making shorter passes. The average yardage per completed pass is falling, while the percentage of passes completed is going up.

In other words, NFL teams are passing more, but they are safer, shorter passes. It helps that some of the current quarterbacks are brilliant – but those that bemoan the demise of the running game needn't be quite so down.

Aim high

NFL quarterbacks with 4,000+ yards in a season

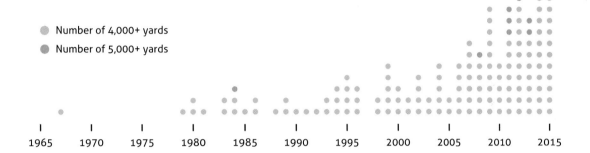

● Number of 4,000+ yards
● Number of 5,000+ yards

1965 1970 1975 1980 1985 1990 1995 2000 2005 2010 2015

Has passing taken over the NFL?

Passing attempts and passing yards as a percentage of total

● Pass attempts (% of plays)
● Passing yards (% of total)

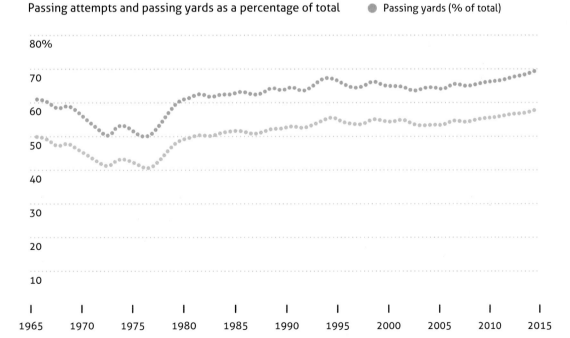

80%
70
60
50
40
30
20
10

1965 1970 1975 1980 1985 1990 1995 2000 2005 2010 2015

What 2-point conversions tell us about the NFL and risk

There comes a time to ignore the stats and play the game

What is the riskiest thing you can do in American football? There are several contenders, but one is the 2-point conversion.

For those unfamiliar with the rules, it goes like this: a team scores a touchdown by getting the ball over the line into their opponents' endzone, which scores six points. They can then opt for a conversion kick from two yards out to get an extra point, which in recent seasons was a 99 per cent guarantee.

Or, the team can try and get the ball over the line a second time by passing or running like a normal play, to get an extra two points. But this is risky, as it's much harder to do.

The choice depends on the scoreline: a two point conversion (known by the acronym 2PAT, for point-after-touchdown) is the obvious thing to do if, at the end of a game, a team trails by 8 points and then scores a touchdown. Losing by 1 point is senseless. You might as well go for it.

Too close to perfect

PAT and 2PAT success rate and attempt rate the following season

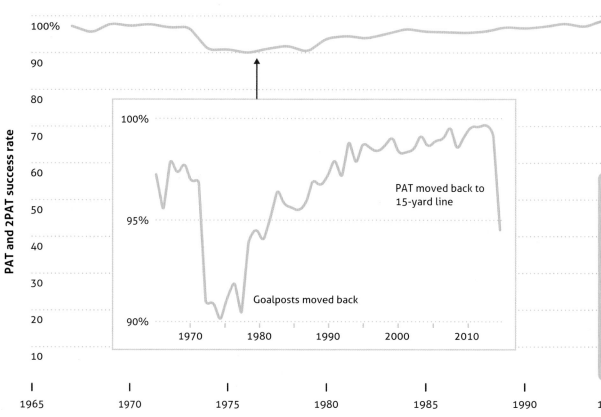

But if you trail by 7, a tie is far the safer option. Equally, trail by 9 or 10 points and a 7 point TD is the best option, assuming there's time for one more drive and a potential field goal, which scores 3 points.

When the 2PAT was introduced into the NFL in 1994, there was a rush of enthusiasm: over 116 were attempted that year, which works out at a rate of 11 per cent. Since then, the attempts have tailed off, with only 41 in 2006 – that's a two-point attempt following only 3.5 per cent of the touchdowns.

In the first few years the success rate fluctuated between 40 and 50 per cent. And teams seemed to pay close attention: as the chart shows, the two-point attempt rate tracked the success rate very closely – if we take the attempt rate the following year, it follows the success rate in step until 2001.

What is interesting is that at a certain point coaches seemed to have a change of heart – somewhere around 2002/2003. From then on, the 2PAT attempt rate plunges below 8 per cent, while the success rate sticks at the 40 to 50 per cent range.

Why? Coaches got more experienced at how to use the two-pointer. They stopped thinking 'last season this worked' and paid more attention to the game.

As a strategy, the 2PAT is clearly only useful some of the time. It makes sense that the attempt rate should settle at around 4 to 6 per cent. There are on average around five touchdowns per game. Let's say every four games or so the scoreline is close enough at the end of the game to make a 2PAT worth going for, so that would generate a 2PAT rate of around 5 per cent.

Meanwhile, the increasing success rate of the PAT, above 99 per cent for the last four seasons, prompted the league to move the position the kick is taken from back to the 15-yard line, from the 2-yard line where it has been for generations. The last time such a big change was made, in 1974 as the goals were moved back from the goal line to the end line, the PAT rate dropped to 90 per cent, and took over a decade to recover.

The 2015–16 regular season saw an immediate effect: the 1-point PAT rate fell from over 99 per cent to 94.2, the lowest since 1982. The 2-point attempt rate went back up, to over 7 per cent of touchdowns scored, back to the 2002 level. In other words, the PAT became less certain, so the 2PAT became more tempting.

What happens next? The 1974 change would suggest that the kickers will get used to the change, become more accurate and the PAT success rate will go back up. That would make the 2PAT less appealing – unless, of course, the 2PAT success rate changes, but that seems stuck in the 45–50 per cent range.

A new equilibrium of risk and reward will be established, until the league tinkers again, probably in another twenty years or so.

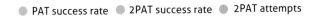

● PAT success rate   ● 2PAT success rate   ● 2PAT attempts

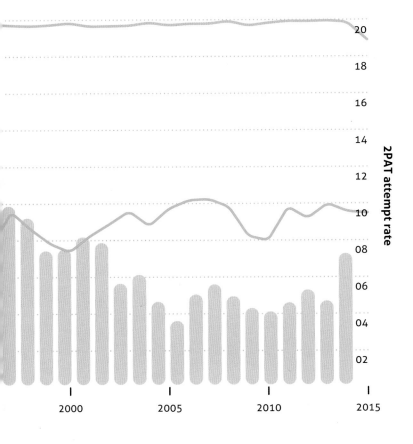

2PAT attempt rate

20
18
16
14
12
10
08
06
04
02

2000   2005   2010   2015

NFL lessons in geography

The divisional map makes little sense; London even less

A look at the map of the NFL by division (the groups which all play each other home and away every season) shows some rather loose definitions of the points of the compass. Indianapolis is by no stretch of the imagination a team in the south of America. Nor is Dallas anywhere near the east of the country.

An accident of reorganisations and relocations, as well as efforts to maintain old rivalries, has made the geography of the NFL rather confusing. Things could be worse by 2022. That's the NFL's target year for a team based in London.

Regardless of which division the London team joins (assuming it all goes to plan), it will skew the distances teams have to travel enormously. Previously, the team that travelled the most on a regular basis was the St Louis Rams, with a combined 7,600km to reach all of their three divisional rivals. As of 2016, the Rams will be based in Los Angeles – the first NFL team back in the US's second biggest city in over 20 years – and their divisional travel distance will fall to 2,760km. The team with the biggest travel commitments to play their division will then be the Dallas Cowboys, who are a combined 6,300km from their rivals.

Of course, the 10 other regular season non-divisional fixtures can throw up some big travelling commitments, especially if the Seattle Seahawks have to play the Miami Dolphins, over 4,400km away, as they did in 2012.

London though, is another dimension: at least 5,000km from Boston, the nearest NFL team, and nearly 9,000km from San Diego.

Given that the NFL has increased the quota of games at Wembley stadium to three per season, and is attracting sell-out crowds, this is clearly the way the NFL wants to expand: overseas markets. First London, then Germany, or Brazil? NFL fans in St Louis are now without a team. Which city will be next?

NFL lessons in geography

Seattle

Oakland Raiders

San Francisco 49ers

LA Rams

San Diego Chargers

Arizona Cardinals

● NFC East	● AFC East
● NFC North	● AFC North
● NFC South	● AFC South
● NFC West	● AFC West

······· The St Louis Rams relocated to LA in 2015

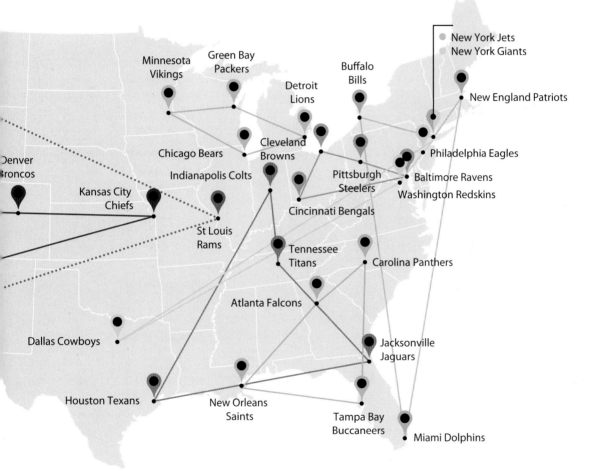

Minnesota
Vikings

Green Bay
Packers

Detroit
Lions

Buffalo
Bills

New York Jets
New York Giants

New England Patriots

Denver
Broncos

Chicago Bears

Cleveland
Browns

Philadelphia Eagles

Indianapolis Colts

Pittsburgh
Steelers

Baltimore Ravens

Kansas City
Chiefs

Washington Redskins

St Louis
Rams

Cincinnati Bengals

Tennessee
Titans

Carolina Panthers

Atlanta Falcons

Dallas Cowboys

Jacksonville
Jaguars

Houston Texans

New Orleans
Saints

Tampa Bay
Buccaneers

Miami Dolphins

on

The case for college football

NFL vs. NCAA: who gets the bigger crowds?

The NFL boasts of being the best-attended professional sporting league in the world. And it has a point: at over 60,000 average attendance per game, it is way ahead of the European football leagues, as well as baseball, or any other major sport.

But what about college football? It might not be professional, but it's certainly popular. It attracts some hefty crowds: the biggest teams regularly get over 100,000 per game. So why isn't the average attendance higher than the NFL's?

The reason is that the top division of college football is comprised of 125 teams. And unsurprisingly, there aren't 125 gigantic stadiums – no league could support that. Across what is called Division I-FBS (the top division), the average attendance for college football in 2015 was 44,000 – which is still a very impressive number given that it's higher than Germany's Bundesliga (43,500, 2014–15), England's Premier League (36,000, 2014–15) and any other professional league, according to Sportingintelligence.com.

However, this isn't comparing like with like. If we compare the top 32 NCAA (National Collegiate Athletic Association) college football teams' attendance to the 32 teams of the NFL, which league comes top?*

It's no contest. In 2015, the college teams attracted an average crowd of 81,000, compared to the NFL average of just over 68,500.

Ah but ... the college teams have bigger stadiums. If the NFL had the same sized venues, would that even things up? Not necessarily. While some teams are full to capacity in the NFL, others struggle to fill the seats. In fact, both leagues have very similar overall capacity numbers: 96 per cent (NFL) and 97 per cent (college) overall capacity in the 2015 season.

There are a few things to bear in mind: college teams are often in areas with no NFL team, so get a strong local following as well as the student fans. College games are also often much cheaper than the NFL. The large capacity is partly a historical accident, and partly because the stadiums are often not all-seater arenas, unlike the NFL.

But caveats aside: far from the NFL being the biggest league in the world, in terms of attendance it's not even the biggest league in its own sport in its own country.

*There are 32 teams in the NFL, but the NCAA figures only give attendance figures for the top 30 games.

NFL team	10,000	20,
Dallas		
NY Giants		
Green Bay		
NY Jets		
Denver		
Washington		
Kansas City		
Carolina		
New Orleans		
Houston		
Baltimore		
San Francisco		
Atlanta		
Buffalo		
Philadelphia		
Seattle		
New England		
San Diego		
Cleveland		
Indianapolis		
Miami		
Pittsburgh		
Arizona		
Tennessee		
Chicago		
Tampa Bay		
Jacksonville		
Cincinnati		
Detroit		
Oakland		

NFL vs. college football: crowded out

Average attendance at the top-30 games in 2015

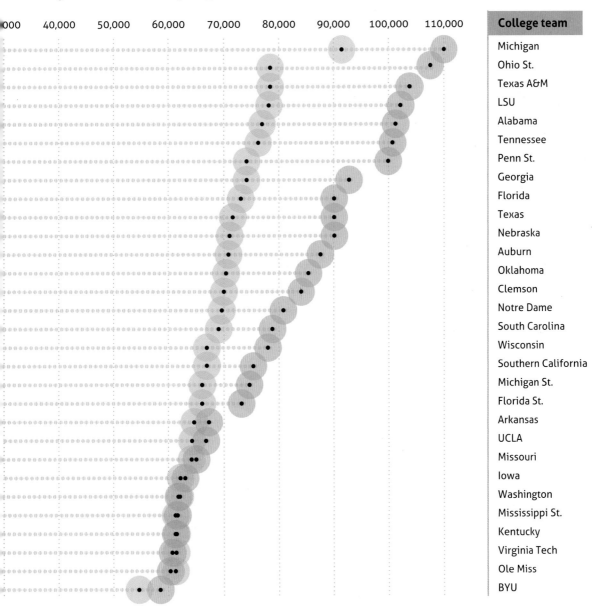

000	40,000	50,000	60,000	70,000	80,000	90,000	100,000	110,000

College team

Michigan
Ohio St.
Texas A&M
LSU
Alabama
Tennessee
Penn St.
Georgia
Florida
Texas
Nebraska
Auburn
Oklahoma
Clemson
Notre Dame
South Carolina
Wisconsin
Southern California
Michigan St.
Florida St.
Arkansas
UCLA
Missouri
Iowa
Washington
Mississippi St.
Kentucky
Virginia Tech
Ole Miss
BYU

An anatomy of 'The Drive'

How the greatest 5 minutes in NFL history unfolded

Legendary quarterback John Elway was one of the best-ever comeback specialists, frequently rallying his team, the Denver Broncos, to fourth-quarter victories.

His most famous moment was during the 1986 American Football Conference Championship Game against the Cleveland Browns, played in January 1987. The Pro Football Hall of Fame described it as the 'prototype performance in the clutch'.

The Broncos trailed the Browns by seven points in the fourth quarter, with 5:32 left on the clock. After dithering on the kickoff, they had the ball on their own 2-yard line, 98 yards from the touchdown required to tie the game. As Elway said in an interview later, it was 'do or die'. What followed is now simply known as 'The Drive'.

Play by play
The Drive markings

1 – First and 10, Denver 2-yard line (5:32 remaining). Elway fakes a handoff to Gerald Willhite, and then passes to Sammy Winder for five yards.

2 – Second and 5, Denver 7-yard line. Winder runs for three yards.

3 – Third and 2, Denver 10-yard line. Denver calls time out. Winder runs over left guard for two-yard gain. First down.

4 – First and 10, Denver 12-yard line (4:11 remaining). Winder runs for three yards.

5 – Second and seven, Denver 15-yard line (3:23 remaining). Elway forced out of the pocket, scrambles for 11 yards. First down.

6 – First and ten, Denver 26-yard line. Elway completes 22-yard pass to Steve Sewell. First down.

7 – First and ten, Denver 48-yard line (2:32 remaining). Elway completes 12-yard pass to Steve Watson. First down.

8 – First and ten, Cleveland 40-yard line (1:59 remaining). Elway pass intended for Vance Johnson is incomplete.

9 – Second and ten, Cleveland 40-yard line (1:52 remaining). Elway sacked for 8-yard loss. Cleveland celebrates.

10 – Third and 18, Cleveland 48-yard line (1:47 remaining). Elway completes 20-yard pass to Mark Jackson. First down.

11 – First and 10, Cleveland 28-yard line (1:19 remaining). Elway pass intended for Watson is incomplete.

12 – Second and 10, Cleveland 28-yard line (1:10 remaining). Elway completes 14-yard pass to Steve Sewell, who fails to get out of play. The clock remains ticking. First down.

13 – First and ten, Cleveland 14-yard line (0:57 remaining). Elway pass intended for Watson is incomplete.

14 – Second and 10, Cleveland 14-yard line (0:42 remaining). Elway forced out of the pocket, scrambles for 9 yards and gets into touch. Clock stops.

15 – Third and one, Cleveland 5-yard line (0:39 remaining). Elway completes 5-yard pass to Jackson for the touchdown. Kicker Rich Karlis adds the point after.

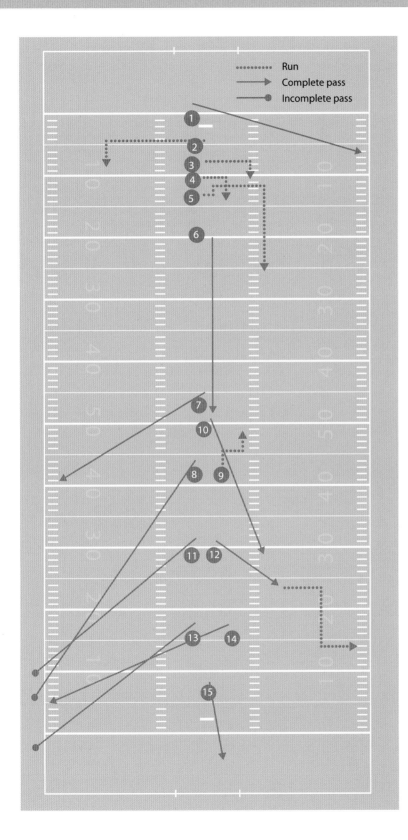

Run
Complete pass
Incomplete pass

The score was now tied at 20–20 with 31 seconds remaining in the game. On Denver's first possession in overtime, the Broncos advance 60 yards in nine plays and the game ends with a 33-yard field goal by Karlis with 5:48 elapsed.

The Broncos went to Super Bowl XXI where they lost to the New York Giants, 39–20. After two further Super Bowl losses in the following three seasons, Elway and the Broncos finally won in the 1998 season, and followed up with another Super Bowl victory the next year. That game, Super Bowl XXXIII, was Elway's last as a professional.

Why stadiums are getting smaller

And more expensive

When it comes to buildings, it's hard not to be impressed by size. The skylines of New York, Hong Kong and now Dubai are rightly held in awe, yet when it comes to stadiums, size isn't everything. In fact, it almost seems like it has become a secondary consideration.

Which is odd, really, when you look at other construction projects. Shopping centres are boasted about in terms of square metres. The desire to build upwards is stronger than ever. No sooner had the Burj Khalifa been completed in Dubai than plans for a taller, 1km high tower in Saudi Arabia were announced. Of the 100 tallest buildings under construction, China is building 63 – and nine of them are over 500m, which only a decade ago would have made any of them the tallest building in the world.

Not so with sport. We don't boast about stadium size any more. The number of spectators that can fit into an arena isn't part of the arms race. In fact, stadiums are getting smaller.

In a sense, the list of largest stadiums by capacity (see chart) can be viewed as a proxy for a bygone age of American grandeur and public pride. Most of the 90,000-plus venues are in the US and were built in the halcyon days of the 1920s and 1930s. They are used for American football, but none in the list is used in the professional NFL. Instead, they are college venues – and amazingly many are usually near-capacity game after game.

In terms of modern sports, this is an anomaly. Building a new stadium is now about competing on the quality of the facilities, the view, the ease of access. Having more seats than your rivals isn't a guarantee of atmosphere, and it certainly doesn't mean a better view for the fans at the back, whom you want to charge as much as possible.

None of this is applicable to the largest stadium in the

Are stadiums getting smaller?

Stadiums with a capacity of over 80,000

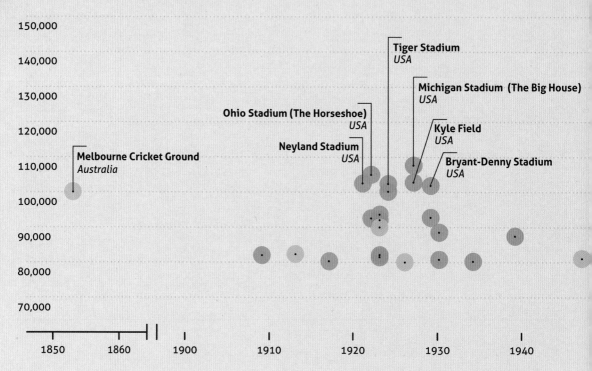

world, though. The Rungnado May Day Stadium in North Korea seats 150,000 but isn't subject to market forces like others. Perhaps it's no coincidence that the world's biggest stadium is in a country where complaining about cost or transport or the facilities isn't an option.

Everywhere else, the trend is towards a different economic model. Sports events have moved from being a high-volume, low-price experience to a lower-volume, premium event product, with exclusive suites, conference facilities, and TV rights on top. That means, of course, the product must deliver – which means state-of-the-art stadiums, and (hopefully) top quality players. All of which is very expensive.

Reliable data is hard to come by, as developers and owners change their designs, but there are only two 80,000 stadiums being planned or built at present – the Hangzhou Sports Park Stadium in China and the Grand Stade de Rugby just south of Paris.

This all points to a future of smaller stadiums, especially in the US. Several baseball and NFL owners in the US have admitted as much in public. In the UK, the top football

teams are looking to expand, but 60,000 is around the limit. Despite a reduction in size, there certainly isn't a reduction in cost. State-of-the-art doesn't come cheap, and several stadiums have broken the $1bn cost price tag. The first was Wembley, in 2007, which cost around £800m ($1.4bn inflation adjusted); the new Dallas Cowboys' AT&T Stadium cost $1.3bn and opened two years later. Since then, New York has led the way in spending big.

The 2009 new Yankee Stadium cost $1.5bn. The same year baseball's Mets moved into their new ground, the $900m Citi Field in Queens. The following year the new NFL arena for the Giants and the Jets – the MetLife stadium – opened costing $1.6bn, the most costly stadium ever. But in a way, that's nothing compared to the renovation costs of Madison Square Garden, home of the Knicks basketball team and others, which when adjusted for inflation has totalled over $1.1bn for a venue of less than 20,000 in capacity. That's over $55,000 per seat, far more than the other New York venues – or anywhere else, for that matter.

For a century, the US led the way in big stadiums. Now it leads the way in expensive ones.

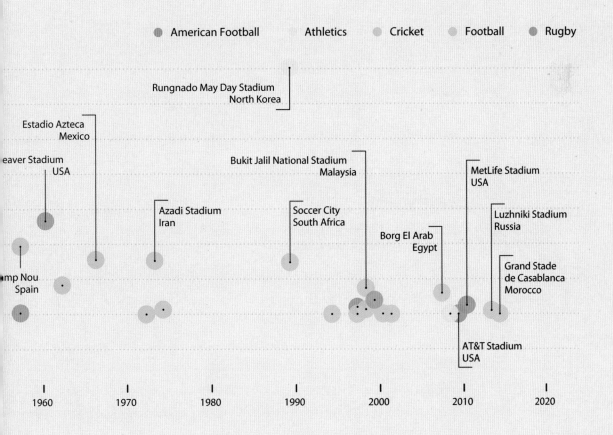

● American Football   ● Athletics   ● Cricket   ● Football   ● Rugby

Rungnado May Day Stadium
North Korea

Estadio Azteca
Mexico

eaver Stadium
USA

Bukit Jalil National Stadium
Malaysia

MetLife Stadium
USA

Azadi Stadium
Iran

Soccer City
South Africa

Luzhniki Stadium
Russia

Borg El Arab
Egypt

Grand Stade
de Casablanca
Morocco

mp Nou
Spain

AT&T Stadium
USA

1960   1970   1980   1990   2000   2010   2020

All-rounders need time to prove their worth

The cricket all-rounder is a wonderful, yet tricky concept

Cricketers are usually batsmen or bowlers. And fair enough: being selected for international cricket for one of those skills is hard as it is. There are however a select few players who truly were Test cricket all-rounders.

One frequently-used way to identify great all-rounders is a career of over 3,000 runs and 200 wickets. This is not easy: only 12 players have ever achieved it. However, this list has

players that definitely fall into one specialism or other. For example, Shane Warne. The great leg-spinner was never an all-rounder, but due to his long career and decent tail-end batting, he got 3,154 runs (along with 708 wickets).

So how to exclude those who meet the wickets and runs criteria, but fail the all-rounder 'smell test'?

Let's look at averages. The average of runs conceded per

Give them time

Batting average vs. bowling average over Test matches played

● Kallis   ● Botham
● Sobers   ● Kapil Dev

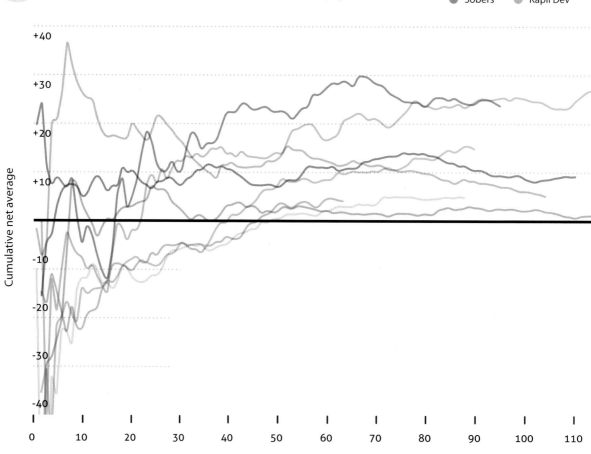

wicket taken tells you quite a lot about a bowler. A world-class average is below 30. A player's batting average shows their contribution per innings, and world-class batsmen average over 40.

If an all-rounder has a batting average higher than their bowling average, their team contribution is clearly positive. Those with a bowling average higher than batting may still be worth a place if either their bowling or batting is consistent, and they have the occasional big match in the other area of the game – such as Andrew Flintoff, who clearly merited a place as an all-rounder later in his career, but never quite truly excelled at batting.

So, if we remove from our list of 12 those with a

⬤ Imran Khan   ⬤ Pollock
⬤ Hadlee   ⬤ Cairns

| | | | | |
|130|140|150|160|170|

combined career average deficit, we are left with eight players: Jacques Kallis (South Africa), Sir Gary Sobers (West Indies), Kapil Dev (India), Sir Ian Botham (England), Imran Khan (Pakistan), Shaun Pollock (SA), Chris Cairns (New Zealand) and Sir Richard Hadlee (NZ).

Pollock and Hadlee clearly fall into the category of bowlers who were decent batsmen. Tellingly, they both only got two centuries, but over 400 wickets.

At the other end of the spectrum, Sobers and Kallis were clearly outstanding batsmen who could bowl well. Their run records are stunning: over 8,000 for Sobers with 26 centuries, 13,000 for Kallis with 45 centuries; they both averaged over 50. In comparison, their bowling was good, but clearly a second skill. They took over 200 wickets, but neither took five wickets in an innings more than six times.

Which leaves Botham, Kapil Dev, Imran Khan and Cairns. All four players have a more equal record between batting and bowling. Botham and Kapil Dev scored over 5,000 runs and took around 400 wickets. Cairns had a shorter career, but was equally effective with bat and ball.

Arguing who is the best all-rounder of all time misses the point. There is a more revealing aspect to the all-rounder.

If we look at the net averages (batting minus bowling) of our eight all-rounders over their careers, what is evident is that it is worth persisting with all-rounders. Players now get dropped quickly at Test level if they don't perform. Yet most of the all-rounders in our list took time to have a positive impact. Kallis took 22 Tests before his net average was above zero. Sobers was positive from his 17th Test to the end of his career. Imran Khan, Hadlee and Cairns all took around 40 Tests before their net average was positive.

Generally, all-rounders improve with age. Perhaps one reason there are so few currently playing is because they are not being given time. Only Botham had a different trajectory, with an exceptionally high net average peaking after just seven Tests.

And all-rounders have another, immeasurable worth aside from just numbers: they inspire others around them, crowd and teammates alike. There is hope: Bangladesh's Shakib Al Hasan fits the bill, and England's Ben Stokes. Just give them time.

Recounting the centuries

What if we counted 100s differently in cricket?

Cricket is a strange game. Why do batsmen get a '1' in the 100s column whether they score 101 or 301? It seems unfair. Instead, if we count a double century as 2, a triple as 3, and (for Brian Lara of the West Indies) 400 not out as 4, let's see how the list of top century-makers changes.

Australian great Don Bradman's total goes up by nearly half again, from 29 to 43, moving him from 13th to 6th in the ranking. Sri Lankan Kumar Sangakkara and Brian Lara don't move much in the pecking order, but both add 12 to their total, giving a truer reflection of their careers. Other players don't add much as they rarely went over 200 in an innings: for instance, Steve Waugh gets a measly extra one to his total, the lowest of any player on the list.

Clearly some batsmen are being short-changed in the all-time century-makers list. Sachin Tendulkar of India stays on top, but lower down the list there are some big jumps. Wally Hammond of England gains eight centuries, and leaps 10 places in the ranking, the largest jump on the chart.

Lost Centuries

● Official 100s   ░ Extra 100s

The change in total number of centuries by player, if centuries were scored differently

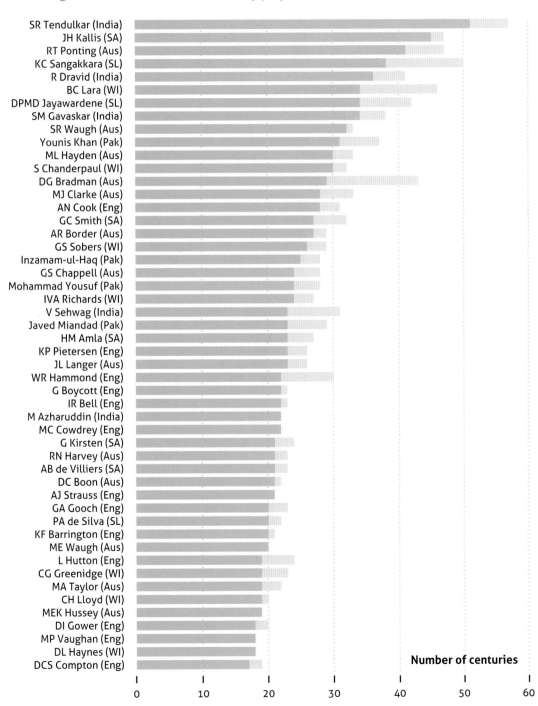

SR Tendulkar (India)
JH Kallis (SA)
RT Ponting (Aus)
KC Sangakkara (SL)
R Dravid (India)
BC Lara (WI)
DPMD Jayawardene (SL)
SM Gavaskar (India)
SR Waugh (Aus)
Younis Khan (Pak)
ML Hayden (Aus)
S Chanderpaul (WI)
DG Bradman (Aus)
MJ Clarke (Aus)
AN Cook (Eng)
GC Smith (SA)
AR Border (Aus)
GS Sobers (WI)
Inzamam-ul-Haq (Pak)
GS Chappell (Aus)
Mohammad Yousuf (Pak)
IVA Richards (WI)
V Sehwag (India)
Javed Miandad (Pak)
HM Amla (SA)
KP Pietersen (Eng)
JL Langer (Aus)
WR Hammond (Eng)
G Boycott (Eng)
IR Bell (Eng)
M Azharuddin (India)
MC Cowdrey (Eng)
G Kirsten (SA)
RN Harvey (Aus)
AB de Villiers (SA)
DC Boon (Aus)
AJ Strauss (Eng)
GA Gooch (Eng)
PA de Silva (SL)
KF Barrington (Eng)
ME Waugh (Aus)
L Hutton (Eng)
CG Greenidge (WI)
MA Taylor (Aus)
CH Lloyd (WI)
MEK Hussey (Aus)
DI Gower (Eng)
MP Vaughan (Eng)
DL Haynes (WI)
DCS Compton (Eng)

Number of centuries

0   10   20   30   40   50   60

England and the ODI fallacy

England DO play enough one-day cricket, they just don't play it well enough

England don't play enough one-day cricket to challenge for the World Cup. True?

It was true. It's not any more, yet it's a myth that's still doing the rounds. A recent version comes from retired Sri Lankan spin bowling legend Muttiah Muralitharan. In an interview with the *Telegraph*, he said: 'they [England] don't play enough one-day internationals abroad like other countries. We play 30–35 one-day matches in a year but England play about 14 or 15, so that is not enough. They play more Tests and that is why they are good in Tests, but they think domestic matches are enough to experience one-day cricket. It is not.'

Murali is spot on, if you look at 2012. That year, England played 15 one-day internationals (ODIs) compared to Sri Lanka's 33.

That year, however, is something of a blip in the last decade or so. In 2007, England (34) played more ODIs than Sri Lanka (29). It was the same in 2011 – England played 30, Sri Lanka 28. In other years, it's been Sri Lanka that's played more, but not by much.

Of course, it was very much the case that England played far too few ODIs to be competitive. In cricket, just as any other sport, results are key. But you also need experience, and simply being less exposed to the 50-over version of

Play more, win more?

One-day internationals per major team, 4-year average

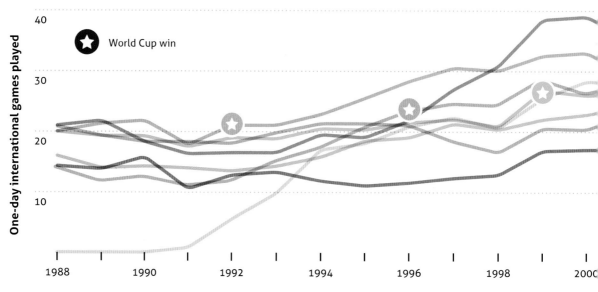

★ World Cup win

One-day international games played

40

30

20

10

1988   1990   1992   1994   1996   1998   2000

international cricket is going to hamper you when it comes to performing at the World Cup.

From 1992 to 2002, England played the fewest ODIs of any of the major cricket sides (excluding Bangladesh or Zimbabwe in this analysis).

Think about this from a player's point of view. If you had started in the England team in 1992 (and didn't get dropped or injured), it would have taken you until somewhere in 2004 to get 200 ODI matches under your belt. If you had played for India, your 200th cap would have been in 1999, a full 5 years earlier. India play the most ODIs, so let's compare them with other teams. The next slowest to 200 caps would have been from New Zealand and the West Indies, with their equivalent players getting their 200th cap in 2001, a good 3 years earlier than our English player.

Since 2003, that's all changed. In the 12 years since 2003, England have played far more one-day cricket, more than South Africa, New Zealand, and the West Indies, and not far behind the others. In that time, an ever-present player would have picked up 273 caps; an Indian player could have accumulated 350 caps. The gap has narrowed.

The four-year rolling chart of ODIs played says it all. England scrape along the bottom until 2003, but have picked up since and play a comparable number of ODIs to the other major teams. The rolling four-year chart is better than the year-by-year version which has too many ups and downs to illustrate the trend of matches played. Four years is also the normal gap between World Cups.

Of course, the teams that play the most don't always win the World Cup: Australia have shown that. But in the case of Pakistan, India and Sri Lanka, as well as the 2007 Australians, their World Cup wins came after a four-year cycle of being one of the top two teams in terms of matches played. Experience isn't everything, but it clearly helps.

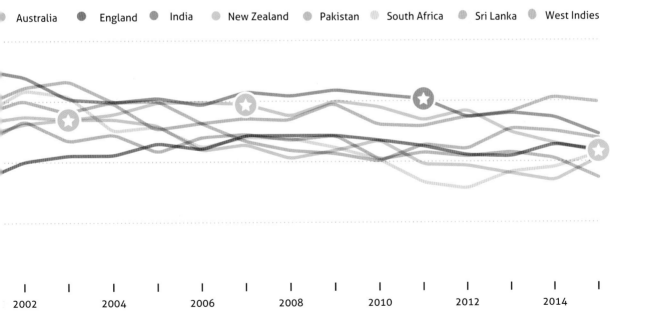

Australia   England   India   New Zealand   Pakistan   South Africa   Sri Lanka   West Indies

2002   2004   2006   2008   2010   2012   2014

Test cricket and the battling draw

Have modern batsmen lost the art of survival?

There is an old saying: cricket is the only sport that can last for five days and still end up in a draw.

Not so much these days. The draw in Test cricket is dying out. From the mid-1960s to the mid-1990s, when the great Australian era began, around four or five out of ten Tests ended in a draw. Then the rate began to drop, and has levelled out at around one-quarter of Test matches.

Why is this? The run rate per over tells part of the story: teams have aimed to score at a faster rate. Instead of scoring slowly to build an innings, teams have looked to build up a score more quickly to give themselves enough time to get the 20 wickets needed to win the match. This attacking mindset has been further enhanced by the advent of T20, which has encouraged fast-scoring batsman in all versions of the game.

Wins in Test cricket aren't everything. A narrow escape with a draw can shift the momentum in a Test series. But with more decisive Tests, has the art of survival been lost? Are teams less capable of hanging on for a draw, as many commentators suggest when a team (often England) is meekly bowled out?

Apparently not. While the increased win-rate may well be a symptom of teams capitulating rather than fighting for a lost cause, the numbers would suggest that the battling draw is alive and well.

If we take a battling draw to be any drawn Test with a fourth innings of over 40 overs – crudely, enough time to bowl a team out – then the number of battling draws as a proportion of all Tests has stayed around the 10 per cent mark since the late 1980s. Since 2004–05, when T20 began, it has actually climbed up a bit. In other words, inconsequential draws are becoming less common – which should be good for Test cricket.

The criteria for a battling draw aren't perfect – this doesn't capture Tests where teams hang on for a draw in the third innings, narrowly avoiding an innings defeat, such as the Ashes Test in Cardiff in 2009. But it does show that the supposed lack of discipline in the new era of fast-scoring batting might be over-stated.

Test cricket: faster scoring, fewer draws

● Draws per test   ● Battling draws per test   ● Run rate

Drugs in sport: a timeline

Catch me if you can

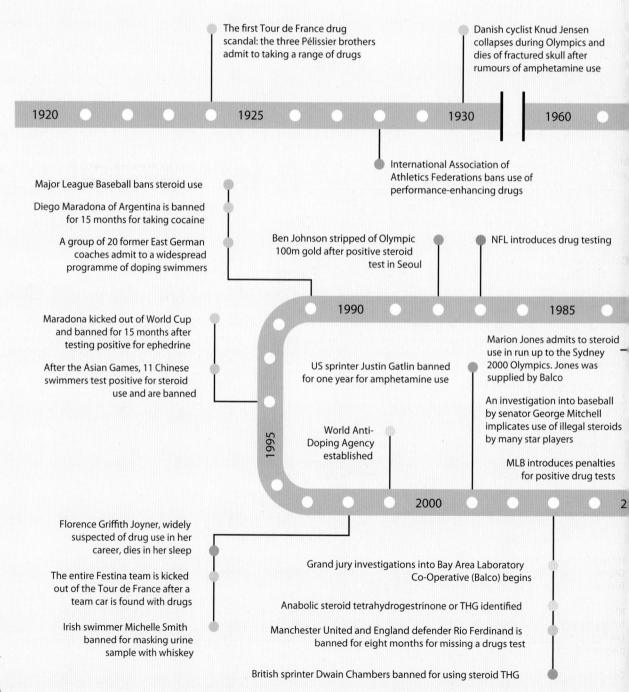

The first Tour de France drug scandal: the three Pélissier brothers admit to taking a range of drugs

Danish cyclist Knud Jensen collapses during Olympics and dies of fractured skull after rumours of amphetamine use

1920 — 1925 — 1930 — 1960

International Association of Athletics Federations bans use of performance-enhancing drugs

Major League Baseball bans steroid use

Diego Maradona of Argentina is banned for 15 months for taking cocaine

A group of 20 former East German coaches admit to a widespread programme of doping swimmers

Ben Johnson stripped of Olympic 100m gold after positive steroid test in Seoul

NFL introduces drug testing

Maradona kicked out of World Cup and banned for 15 months after testing positive for ephedrine

After the Asian Games, 11 Chinese swimmers test positive for steroid use and are banned

1990 — 1985

US sprinter Justin Gatlin banned for one year for amphetamine use

Marion Jones admits to steroid use in run up to the Sydney 2000 Olympics. Jones was supplied by Balco

An investigation into baseball by senator George Mitchell implicates use of illegal steroids by many star players

World Anti-Doping Agency established

MLB introduces penalties for positive drug tests

1995

2000 — 2

Florence Griffith Joyner, widely suspected of drug use in her career, dies in her sleep

The entire Festina team is kicked out of the Tour de France after a team car is found with drugs

Irish swimmer Michelle Smith banned for masking urine sample with whiskey

Grand jury investigations into Bay Area Laboratory Co-Operative (Balco) begins

Anabolic steroid tetrahydrogestrinone or THG identified

Manchester United and England defender Rio Ferdinand is banned for eight months for missing a drugs test

British sprinter Dwain Chambers banned for using steroid THG

Legend:
- American Football
- Athletics
- Auto-racing
- Baseball
- Cycling
- Football
- Horse Racing
- Ice Hockey
- Misc
- Olympics
- Swimming

Raymond Poulidor is the first rider to be tested for drugs in the Tour de France

Football and Cycling governing bodies introduce doping tests. Most major sports have testing in place by 1970s

Belgian cyclist Eddy Merckx tests positive for amphetamine. Merckx also tests positive in 1973 and 1977

1965

1970

[British] cyclist Tommy Simpson dies of [exhau]stion during Tour de France after consuming drugs and alcohol

Drug testing at Summer and Winter Olympic Games in Grenoble and Mexico

[Dancer's] Image disqualified from Kentucky Derby for anti-inflammatory drug

A whistleblower reveals blood test files that show widespread doping in athletics

Wada report suggests state-level involvement of doping cover-ups by Russia

Tour de France winner Chris Froome releases his blood data to counter allegations of doping

Athletics governing body IAAF suspends Russia from all international competition

1980

1975

Steroid testing starts at Montreal Games

IAAF suspends seven Russian athletes prior to Beijing games for doping

Barry Bonds indicted by US federal grand jury for perjury over steroid use

Kentucky Derby bans use of steroids

Mark McGwire admits to steroid use during the 1998 season when he hit a record 70 home runs

Roger Clemens indicted by US federal grand jury for lying to Congress over steroid use. He is acquitted in 2012

US cyclist Floyd Landis who was stripped of 2006 Tour de France title after testosterone test admits to doping

2010

2015

2020

Having won the 2004 100m Olympic title, Gatlin is banned for four years for a positive drugs test

NHL player Bryan Berard is banned from international competition for taking steroids, but is allowed to play in NHL

Gene doping banned in US

Lance Armstrong investigation dropped by US Attorney's Office

Alberto Contador, 2010 Tour de France winner, found guilty of taking clenbuterol

Lance Armstrong declines to contest doping charges in US, and is stripped of seven Tour de France titles

Godolphin, the largest horse trainer in the world, is found to have given 11 horses steroids, against UK rules

MLB investigation results in suspension for 12 players including Alex Rodriguez, baseball's highest-paid player

Millionaires everywhere

Golf inequality is falling

We live in a world of growing inequality, but not in professional golf.

The PGA tour lists the prize money for every player since 1980, and a look at the data is startling. Tom Watson, the prize money leader in 1980, won just over $530,000. The tour leader in 1981, Tom Kite, got just $375,000. In 1988, Curtis Strange became the first player to win over $1m in prize money on the PGA tour, with $1.15m. Tiger Woods became the first player over $2m in a season, in 1997. Since 1999, every PGA prize money leader has won over $4m, and some over $10m in a year.

You might think this is an example of 'to the victor go the spoils'. You would be wrong.

In the 1990s, the number of PGA prize money millionaires starts to grow. Then from 1997 onwards, the number of players winning seven-figure sums in a season explodes.

In 12 years, it went from single figures to over 100 in 2008. Since then, around 80 to 100 golfers a year win over a million dollars. In 2015 on the main PGA tour, 102 players won over $1m. Although many are the same as in previous seasons, some are new, and those who earn under a million will very possibly accumulate well into seven figures over the years. That means there's a lot of millionaire golfers.

While the tour leaders now win around 23 times what they did in the 1980s, the average tour winnings is roughly 30 times higher. Of course we all know that averages can be skewed. What about the median prize money winnings – the middle-of-the-road player? Median winnings have gone up even more, by 60 times in our data sample.

This all means that golf is actually getting more equal. Using a standard economic measure of inequality* shows how the increase in prize money has led to a more equal tour, with the score falling from just under 0.7 to around 0.55 (see explanation below). The big fall coincides with the explosion of prize money.

Put another way: the 10th best player on the PGA tour now frequently wins about 4 times as much as the 100th. It used to be more than 6 times.

The PGA isn't the only golf tour – there are players earning seven figure sums on the European tour too, as well as the Champions tour (for over 40s). In the last few years, around 10 to 15 players usually win over $1m on the Champions tour. The European PGA listed 35 players in 2015 earning over €1m.

There are now probably around 150 players per year earning $1m-plus from golf. Of course, golf players have costs. There are caddies, hotels, travel, agents. It's not all pocketed. And as the chart shows, the explosion in winnings has tailed off a bit, mainly due to the global recession: sponsors had less money to throw at golf events.

Overall though, the big money is in the PGA. And however you cut it, golf's biggest tour is getting more equal.

---

*This is called the Gini coefficient, which shows how wealth is distributed among a group. A score of 1 shows perfect inequality – think of a dictator owning everything, and the population nothing – and a score of 0 is perfect equality, where everyone has the same amount. The US has a Gini score of 0.4.

Join the club: golf winnings

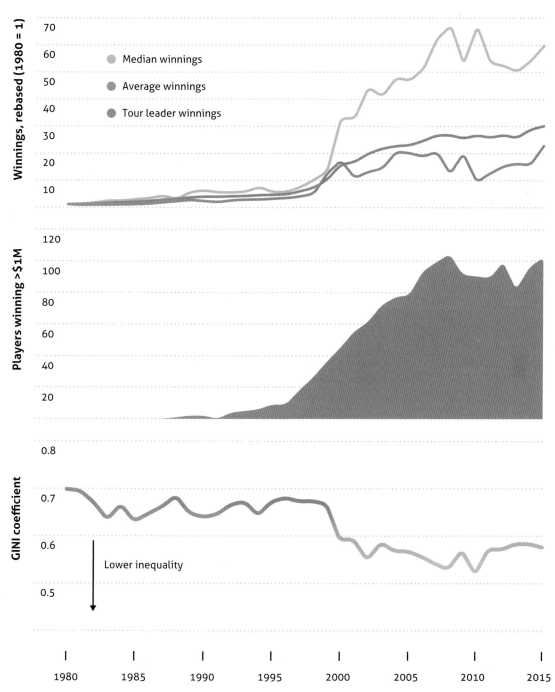

Winnings, rebased (1980 = 1)

70
60
50
40
30
20
10

● Median winnings
● Average winnings
◐ Tour leader winnings

Players winning >$1M

120
100
80
60
40
20

GINI coefficient

0.8
0.7
0.6
0.5

Lower inequality

1980 1985 1990 1995 2000 2005 2010 2015

The short stuff matters

Longer driving hasn't changed golf as much as you might think

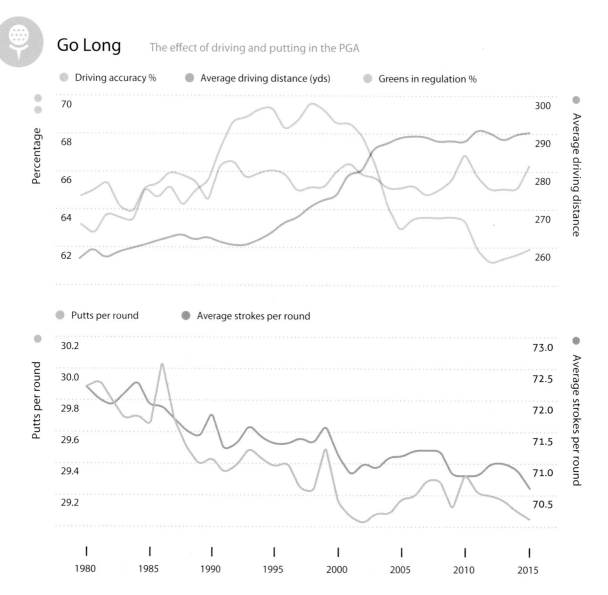

Go Long The effect of driving and putting in the PGA

- Driving accuracy %
- Average driving distance (yds)
- Greens in regulation %

- Putts per round
- Average strokes per round

In 1974, a 64-year-old man picked up a wooden golf club, teed off, and 515 yards later the ball came to a rest. His name was Mike Austin, and it's still regarded by some as the longest golf drive ever.

After all the changes in club technology, such as titanium shafts and grooved club faces, as well as fitter and younger players, how could a senior pro still have the record 40 years later?

It helped that Austin was at altitude, with a strong tailwind, and was a great striker of the ball. The course, near Las Vegas, was hard and bouncy. It was the perfect set up.

Although Austin's mark has lasted, the pros are catching up. In 2015 Chad Campbell hit a tour-leading 489-yard drive. That year, there were 24 drives of 400 yards or more.

Compare that to 20 years earlier. In 1993 there was just one drive over 300 (not 400) yards on the PGA tour. In 2013, there were over 800 drives over 350 yards, and countless more over 300.

The average length of golf drive on the PGA tour didn't change much until 1993. At that point it rocketed up, from an average of 260 yards to around 290 in 2005, mainly owing to huge advances in club technology. Then it hit a plateau. Why? And what has been the impact on overall scores?

LONGER DRIVES AREN'T MORE ACCURATE

One reason for the levelling off is that long drives aren't always an advantage. Hitting long into bunkers or the rough makes for a tougher second shot than being 50 yards back but on the fairway.

So if the pros are hitting within themselves, is driving accuracy increasing? No: from 1998 when the average drive distance was 270 yards, to 2005 when it was 288, the percentage of drives that made it on to the fairway fell from 70 to just over 60, giving up all the gains in accuracy from the previous decade. Did those 18 extra yards make such a difference when the ball went into a trickier position every ten holes or so? That was the trade-off that the PGA players seemed happy with. Was it worth it?

ACCURACY MAKES NO DIFFERENCE FOR GETTING TO THE GREEN

It seems to have paid off: according to PGA data, the rate that golfers get to the greens in regulation – i.e. two shots less than par – hasn't moved much from around 65 per cent. That would suggest the ability to hit a good second shot from the rough has improved. The odd wild drive can be recovered.

PUTTING IS MORE IMPORTANT

Overall, there has been a significant improvement in golf scores since 1980, from over 72 strokes per round to fewer than 71 on the PGA tour. The period when driving average increased the most coincided with a drop in average strokes. But that isn't the most closely correlated improvement. Take a look at the second chart.

The numbers show that it all comes down to putting.

The average number of putts per round tracks the average score per round incredibly closely – a correlation of 0.88 (1 is perfectly correlated). Even if we take out the effect of chipping the ball from near the green to close to the pin, the correlation with putting is still very close (0.84).

So have all these longer drives made the difference? It depends: putting is still key. The long drive only makes a difference if it means an easier second shot (on a par 4 or above), which gives the possibility of getting the ball closer to the pin, which means fewer putts – in theory.

And that seems to stack up: the average driving distance compared to putts per round each year suggests longer drives go with fewer putts.

LONGER DRIVES DON'T GET YOU CLOSER TO THE PIN

It would be even better if we knew how close balls were hit to the pin on the approach shot to the green. And the PGA does provide that data, although only for the last 14 years. In that time, driving distances have increased, and then fluctuated. The hole proximity, as it's called, has fluctuated in a very similar way.

But that's not good: the longer drives have actually coincided with the ball being further from the hole, not nearer as we might expect.

In which case, longer drives often mean less accuracy and result in the ball being further from the pin: what's making the difference to the lower scores? It's back to the putting. And since the early 1990s, the times a player putts just once on a hole has gone up from under 36 to over 39 per cent. That might not sound a lot, but it means around a shot less per round.

In other words: hitting long off the tee might look good, but it's the short stuff that makes the difference.

(It's worth noting that these figures are for the tour in general – i.e. an average of what all the players are doing. To be ranked near the top of the tour, driving is crucial. The players that can hit long drives when required and follow up with the rest of the game stand out. In fact, according to the PGA's new statistic, 'shots gained', which measures relative scores amongst the players, the big difference isn't in the putting, but in tee-to-green. However, shots gained doesn't tell us about the tour as a whole.)

The magic number is 18

How Woods and Spieth took very different paths to the Masters title

Much has been written about the Masters. The organisers are so strict (no mobile phones are allowed) it has been compared to North Korea. It wasn't exactly progressive about admitting black or female members. It's the one of the most beautiful golf courses in the world, yet it is surrounded by deprivation.

It is also a career-defining venue for golfers. Every player wants to don the famous green jacket at Augusta at some point. For two players, that moment came very early in their careers: Tiger Woods in 1997, and Jordan Spieth in 2015.

The similarities are eye-catching; Woods and Spieth were both 21 (Woods was younger by five months); it was their

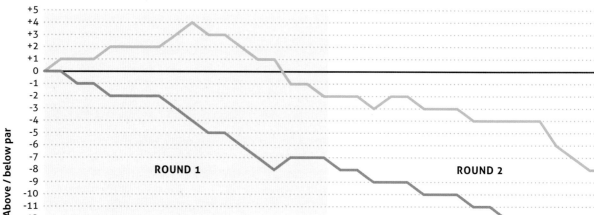

Woods and Spieth: the Masters of Augusta ● Woods (1997) ● Spieth (2015)

first major; and they now share the course record of just 270 shots, 18 below par.

There are differences too. Woods destroyed the competition, winning by 12 shots; Spieth won relatively comfortably by four.

Plotting their scores, hole by hole and round by round, what stands out is just how badly Woods started. After the first nine holes he was four shots over par. After 14 holes, he was nine shots behind where Spieth was. Even after Woods hit an eagle on the 13th followed by two more birdies in his second round, he still was on 8 under at the half-way point,

compared to Spieth on 14 under. It was the third round where Woods played his best, shooting 65 with 7 birdies and no bogeys.

Spieth's Masters was all about his first two rounds, which contained an astounding 15 birdies and just one bogey. His third and fourth rounds were both just two shots below par.

In fact, Spieth got to 18 under towards the end of his third round – Woods wasn't there until a round later. Had Spieth not dropped a shot at the last hole of the tournament, he would have pipped Woods by one stroke. As it is, they are together, frozen in time as 21-year-olds: 18 under, 18 years apart.

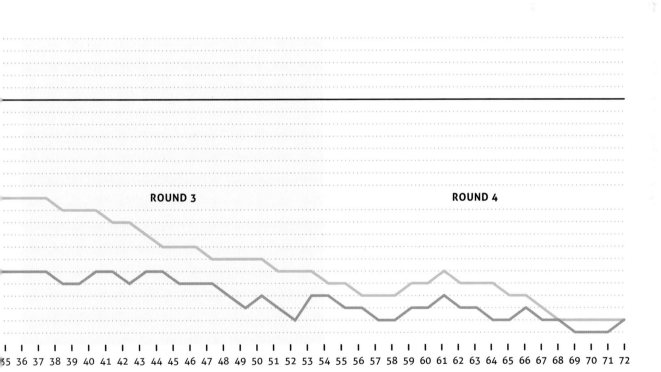

ROUND 3　　　　　　　　　　ROUND 4

55 36 37 38 39 40 41 42 43 44 45 46 47 48 49 50 51 52 53 54 55 56 57 58 59 60 61 62 63 64 65 66 67 68 69 70 71 72

Golf's great global shift?

America's demise and Asia's rise are exaggerated

Over the past few years, hundreds of articles have all sounded the death knell for golf in the US. Meanwhile, Asia is supposed to be where it's at: lots of new courses, lots of newly-rich, keen players. Is this an accurate picture?

It's really a question of supply and demand. The supply is the number of courses available. The demand is the number of people playing. Demand is of course linked to supply: you can't play golf if there aren't any courses.

The main worry is over US demand. From 1986 to 2003, US golf gained over 10m players, totalling over 30m in the mid-2000s. Since then it has shed around 5m of them, according to the National Golf Foundation, with participation falling to 24.7m players.

Asia is a complicated picture* and China is the most complicated place of all. China's newly rich may want to play golf, but from 2015 to 2016 the Chinese Communist Party officially banned golf club membership for Party members and golf course construction, as part of an anti-corruption drive. As a result no senior politician or businessperson would ever admit to playing the sport and there are no reliable figures for how many people play. A recent report estimated there were around 1m golfers in the country, playing at clubs that were often dubbed leisure facilities or country clubs, to help the ruling party turn a blind eye. And yet there is an official PGA China tour. Very confusing.

Japan is still Asia's biggest golf-playing nation, with around 10m golfers. But those numbers have fallen since the 1990s by around 40 per cent. It is likely that over half the world's golfers are still in the US.

What can we learn from the number of courses being built – and shut – worldwide?

It is true that in the US, courses are closing: since 2006, over 100 per year, according to the National Golf Foundation. Many reasons have been put forward: fewer people have time to play golf; the courses are too hard; it's too expensive. So golf in the US is trying to adapt, becoming more family-orientated and female-friendly. Clubs are innovating with quicker formats of the game and becoming less rigid with their rules.

What about Asia? It's impossible to know the exact number of Chinese courses owing to the temporary ban. Many were not listed as golf courses at all to avoid being closed by the authorities. Some reports suggest that over 600 courses have been built in China in the last decade. Then in 2015, 66 courses were shut in the crackdown.

According to the Royal and Ancient, golf's governing body, Japan has no new courses under development, and several recent news reports have highlighted courses being turned into solar energy farms. India is building 32 courses according to the R&A. (India has about 200 registered golf courses but about half of those are on military bases and only accessible to the military.)

In total, the R&A estimate that Asia has 207 courses under development. That is the equivalent of 1.4 per cent of all the courses in the US. Of those 207, less than half are actually under construction.

(At the same time, the US itself has 200 courses under development, according to the R&A, although that doesn't take closures into account.)

The idea that golf has shifted from the US to Europe and Asia overestimates the change in demand, while not taking into account the small shifts in supply.

The dominance of the US is even starker when we take population into account. Using figures from the R&A, I have worked out the number of people per golf facility in each continent, to give an idea of how available golf is. Of course, this doesn't take wealth into consideration, or how popular golf is. Still, it is revealing. Oceania has 18,000 people per golf facility. North America has 31,000, Europe 100,000; South America: 619,000; Asia: 900,000. Finally, Africa has over 1.2m people per golf facility.

In other words, the idea that golf in the US has collapsed while the rest of the world surges ahead is wrong. In terms of number of players and facilities, golf's great shift has hardly begun.

---

*Reliable participation figures for Asia are not widely available, unlike the NGF and EGA, which publish data.

Golf's great global shift

Existing golf facilities

⬤ Africa	⬤ Asia (incl. Middle East)	⬤ Europe
⬤ North America (incl. Central America & Caribbean)		
⬤ Oceania	⬤ South America	

Golf facilities under construction

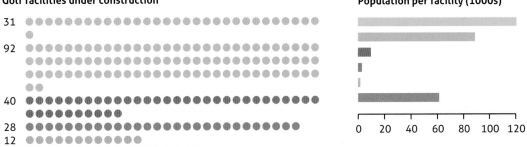

31
92
40
28
12
18

Population per facility (1000s)

0   20   40   60   80   100   120

Where have all the American golfers gone?

The top 100 and the Ryder Cup show the US being pushed out by Europe

With Jordan Spieth the latest new golf sensation winning major titles at a Tiger-esque young age, US golf seems to be in good shape.

The truth is quite different. If top American golfers were wild animals, they would be on an endangered species list. Look at the evidence. First, the top 100 world ranking: once there were 60 US golfers on it. Now there are 40.

In the past decade (2006–15), they have won 28 per cent (18 out of 40) of the major titles. The decade before that it was 70 per cent (28 out of 40). Admittedly, that is helped by Tiger Woods winning quite a lot, but there were many others.

And then there's the Ryder Cup. Once upon a time, it was just a US-British tournament (Britain plus Ireland for three editions). Then in 1979 it was widened to include Europe, after America had won it 18 out of 22 times. It took until 1985 for a European victory, but since then the pendulum has swung.

Europe has won 8 of the last 10 Ryder Cups. Once, the joke was to even things up it should be America vs. the rest of the world. Now, perhaps, the rest of the world should play against Europe. How can we explain the dominance of Europe in the Ryder Cup?

One reason often mentioned is team ethos – that Europeans seem to have a more natural team instinct. This is odd, given that there are almost no other competitions where players represent Europe, rather than their country. The closest comparison would be when the Lions play rugby.

How about preparation and tactics? Europe's 2014 captain Paul McGinley managed the team with great attention to detail, and had learned the ropes as a vice-captain. His counterpart, Tom Watson, hadn't been anywhere near the Ryder Cup for 20 years.

What about location? One theory is that American players play predominantly in the US. European players travel far more, so the US is less alien to them than Europe is to the Americans.

These are all plausible differences. But the clearest marker is talent pool. The top 100 chart shows that US stock has fallen over the years, while Europe's has picked up.

The objection to this argument is that there are just 12 players in a Ryder Cup team, so why would it make any difference if there were 20 or 24 to choose from in the top 100?

Firstly, if you have just 15 players in the top 100, as Europe did in 1987, compared to 59 Americans, it's clear which country has the better talent pool. That has now evened up, and the results show it. Many matches have been close: Europe has won by a single point, the smallest margin of victory, in four of its most recent eight victories. A missed putt here or there and it's a different story.

American golf as a whole is on the decline, with clubs closing and millions of people deserting the sport. In the time that America lost 5 million golf players, from 2003 to 2011, Europe gained a million. But that doesn't necessarily translate to the top 100 golfers in the world.

The answer is simply that at the elite level, Europeans have caught up. The European tour is more competitive, and more Europeans play on the PGA tour as well.

Furthermore, Europe has become more European. That might sound odd, but Europe in the Ryder Cup used to mean the British and Irish players, plus Seve Ballesteros and one or two others from the continent. Now, it's nearly half the team. The first nine European Ryder Cup teams averaged 3.7 players from continental Europe. The last nine have averaged 5.2.

The top 100 shows the same trend: the number of players from continental Europe has grown from a handful in the late 1980s to around 20 now.

It's not so much that the Americans have stopped producing good players, it's that Europe has started. There are only so many spaces in the top 100 to go around.

Balancing out: golf year-end top 100 by region

● United States   ◉ Europe   ● Rest of World   ★ Ryder Cup winner

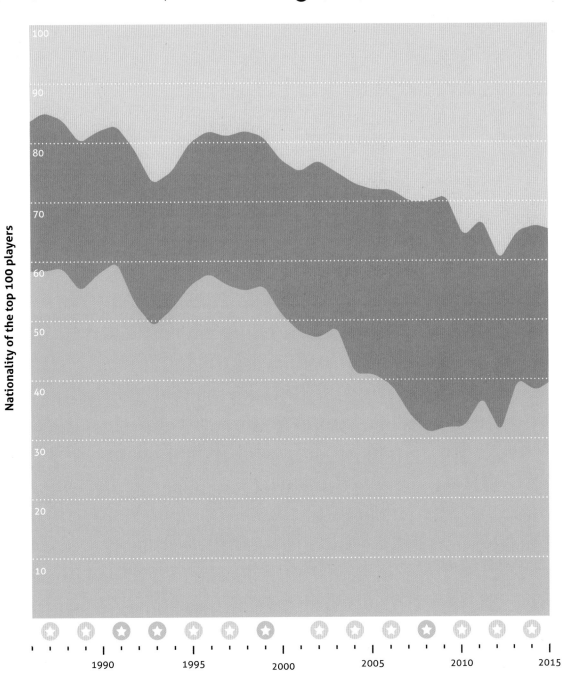

Nationality of the top 100 players

100

90

80

70

60

50

40

30

20

10

1990   1995   2000   2005   2010   2015

Snooker just keeps on getting better

For many sports, comparing standards across generations is hard. Are current footballers that much better than those of the 1990s, or 1960s?

Snooker, however, is a bit different.

Firstly, fitness is largely irrelevant. You could argue that being in good shape helps with long matches, but it's not the fine margins that make the difference in cycling or athletics, for instance.

Second, the equipment and table haven't changed. The balls and surface are largely the same as they were 50 years ago, which isn't true in tennis, or golf. It's a standardised, constant environment.

Thirdly, although players can put their opponent in awkward positions, once a player is onto their second shot, they are alone. Individual skill and psychological pressure are the only factors in play.

Which means snooker is ideal for comparing standards. So which standard to use?

Century breaks are a good measure: that is when a player scores over 100 points in one visit to the table. It's a mark of individual skill and is separate from low-scoring frames that may include brilliant safety play.

Starting in 1982, the year that the first official maximum 147 was made (by Steve Davis), we can see that the number of centuries scored per season by the elite players has shot up, from less than 100 to over 1,000 in recent years.

The number of matches helps this. More matches means more century breaks. But not every match is the same length in terms of how many frames are needed to win. So how do we account for that?

Snooker: The centuries keep going up

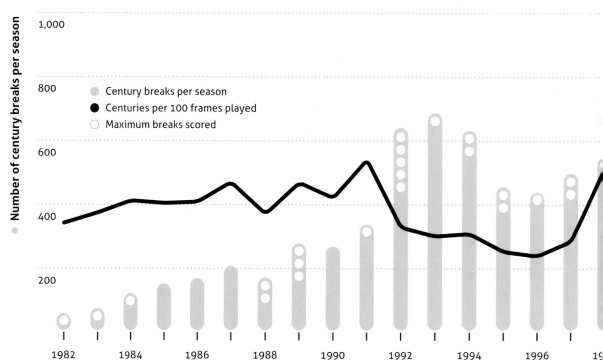

Number of century breaks per season

- Century breaks per season
- Centuries per 100 frames played
- Maximum breaks scored

1,000

800

600

400

200

1982   1984   1986   1988   1990   1992   1994   1996   19

If we look at the total number of frames played, regardless of match, we get a better idea of how frequently players are scoring centuries. And it's remarkable: the number of centuries per 100 frames played has gone from less than 2 per cent to over 6 per cent.

In other words, you could say that snooker players are three times better than they once were.

The number of maximum 147 breaks has also gone up, from one or two a year to often seven or more.

Aside from better players, are there any other reasons for the surge in centuries? For a time there were extravagant bonuses paid to players for getting the highest break in a competition, or for getting a maximum 147. Money is a great motivator. But not every century break begins with enough balls on the table to get a maximum, or a highest tournament break. The higher level of centuries must be to do with better skill.

Furthermore, the rewards have been drastically reduced. A player who shot a 147 at the World Championship a few years back would get £147,000, plus a highest break prize of £18,000. For a few seasons, a maximum break would net you almost the same as winning the whole tournament.

But this was abolished in 2011. Now players get £10,000 for a highest break if televised, and nothing extra for a 147. Instead, there is a rolling 147 fund across major ranking events. Several players have complained, saying that attempting a 147 should have higher rewards, as it often requires riskier shots that might leave an opportunity for the opponent to win the frame.

Which makes the high number of maximums and centuries in the last few years even more interesting: the financial incentive has been reduced, but the players are still knocking them in. Snooker players are simply getting better season after season.

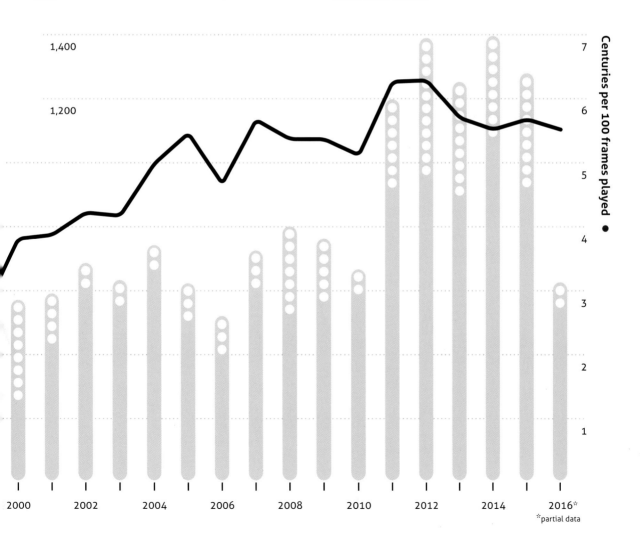

How tennis is killing doubles

Players respond to prize money, and doubles is being undervalued

As the US Open of 2014 came to a close, the victory was described by one of tennis' foremost writers* as 'one of the most impressive feats in the annals of tennis'. Was that Serena Williams winning her 18th Grand Slam crown? Actually, it was the Bryan twins, Bob and Mike, winning their 100th career doubles title.

So doubles is doing well, you might think, spearheaded by the Bryans, a charismatic and amazingly successful pairing. As former player and commentator Mats Wilander put it, 'They have taken it to the next level.'

Doubles is definitely popular at grass roots: go to a set of tennis courts, and chances are you will see as many people

Doubles Down   ○ Singles   ● Doubles

Percentage of Wimbledon prize money to doubles and singles winner

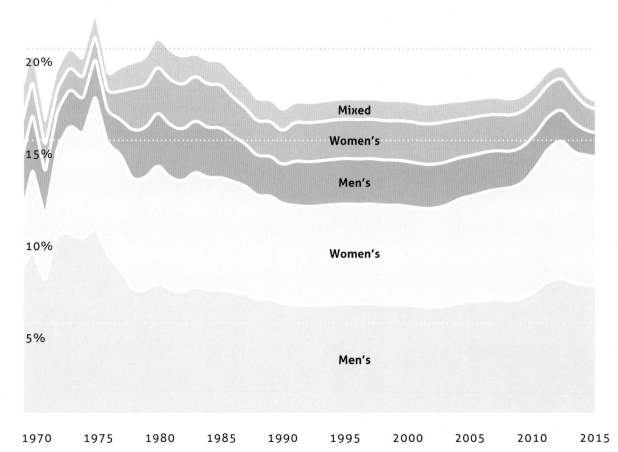

25%

20%

Mixed

Women's

15%

Men's

10%

Women's

5%

Men's

1970   1975   1980   1985   1990   1995   2000   2005   2010   2015

playing doubles as singles. After the Bryan twins, can you name another pairing? Perhaps the Williams sisters, but they have only played doubles at two events outside the majors or the Olympics in the last decade. Jamie Murray, Andy's brother? Can you name his partner?

You probably can't, but whose fault is that? Does tennis fail to market doubles properly, or are the public just not that interested?

Certainly doubles may suffer from a vicious circle of less-interested spectators, which makes promoters push singles far more than doubles, which in turn makes doubles less well known and kills the public's appetite further.

What about the players? In decades past, a doubles match might feature star quality such as John McEnroe, or Martina Navratilova. That has long gone. Almost none of the top singles players play doubles any more, so the two forms of the game are diverging. In the 1995 year-end rankings for men's singles and doubles, there were 38 players that featured in both lists. At the end of 2015, there were just 22 players that were ranked in the top 100 in both forms of the game.

It is even more stark at the top of the tree: in 1995, seven of the top 20 singles players featured in the top 100 of doubles, including four in the top 10 (Boris Becker included). Fast forward 20 years and just one of the top ten had a doubles ranking inside 100 (Rafael Nadal), only three of the top 20. In other recent years, often none of the top 10 have had a doubles ranking in the top 100.

The top singles players only play doubles at special events, such the Olympics or Davis Cup, or when they need court time when returning from injury. The Williams sisters, in case you were wondering, are ranked for doubles outside the top 100, although they were ranked number one in 2010 briefly.

Clearly, the game is more physical now than 20 years ago, and the demands are too great on the top players to play both doubles as well as singles. But what are the incentives (aside from the glory)? When it comes to prize money, the message is clear: doubles just isn't valued any more.

The Bryan twins at the US Open just won $520,000. Sounds good, doesn't it? But per person, their payday of $260,000 is less than a losing quarter-finalist in the singles, who got $370,000. It's closer to the eight players who lost in the fourth round and went home with $187,300 each.

In other words, a player who wins six doubles matches (including the title) is rewarded far less than someone who wins four singles matches and loses the fifth.

It is the same at the other major events. Twenty years ago, a Grand Slam doubles winner was almost on a par with singles semi-finalists. Now they are valued two rounds further back.

Of course, prize money has rocketed up over the years, and it is true that doubles players have seen increases – the winning pair at Wimbledon now receives £325,000, up from less than £200,000 in 2000. But in terms of how the overall pot of money distributed, it's a telling story. Doubles as a proportion of the total prize money is falling – and fast.

Let's use Wimbledon's prize money as a guide. The chart shows the prize money for the winning doubles pair as a percentage of the overall prize money at Wimbledon. It fluctuates at the start of the Open era, from 1968 to 1990. Then in the 1990s it was pretty stable. After 2002, it starts to drop off. And after men and women receive equal prize money in 2007, it plummets. The men's winning doubles pair used to get around 2.5 per cent of the overall pot. At the current rate of decline, it will be less than 1 per cent before 2020.

This is also true of the overall payout to all the players in the doubles event. In 1994, the men's doubles in total received 11 per cent of the overall Wimbledon tournament prize money – just shy of £630,000 of a total £5.7m (the women got £500,000). Twenty years later, and the doubles event received just under 6 per cent of the overall £25m prize money.

Players respond to incentives, and prize money is the ultimate guide.

Doubles is caught in a trap. Less money means fewer stars; fewer stars means lower crowds and TV ratings; fewer ratings means less money. Yet anyone watching the doubles at a Davis Cup tie (it's always the crucial third match of five) would agree that it can produce great drama.

Doubles will always lack the added glamour that singles has. It also places a different emphasis on a player's skills, which in a world of big serves and crunching forehands makes it harder to switch between the two forms of the game. But if tennis tournament organisers want doubles to continue other than as a schedule-filler for purists, a little more prize money might help.

*By Peter Bodo, one the foremost commentators in tennis

From Murray to Perry: 17 degrees of separation

How to connect the British champions over 70 years of matches

'Tennis title home' was the headline in the *Daily Telegraph*, and the simple picture caption in the *Guardian* said it all: 'The Champion'. Every paper celebrated the end of a long wait for a home winner at Wimbledon, the biggest tennis tournament in the world.

Not for Andy Murray in 2013 – it was for Fred Perry in 1934, when he won Wimbledon for the first time. It had been 25 years since the last English winner, AW Gore in 1909. Gore was followed by years of Australian, US and French winners. *The Times* suggested in an editorial that the victory 'will delight the great body of lawn tennis players in this country, and the general public will share their satisfaction.'

The Times also predicted an upturn in British tennis fortunes: 'Perry's success will help lawn tennis in Great Britain because – quite illogically, but none the less inevitably – it will give British players a better conceit of themselves.'

Fast forward 78 years, and Andy Murray's Wimbledon victory was greeted with the same sentiments, just delivered with a greater degree of hysteria: 'Magical Murray'; 'History in his hands'; 'Wimbledone'.

Murray's win was also predicted to give British tennis a boost just as Perry's was supposed to. Tim Henman, for so long the British hope before Murray, said after Murray's victory over Novak Djokovic: 'I hope this victory can be another shot in the arm for British tennis … Andy has inspired a generation.'

Other similarities exist between the two finals. Back in 1934, the *Daily Telegraph* described the crowd as: 'an expectant throng, hoping, but not quite sure until the very end' – that could have equally applied to the hugely tense Murray-Djokovic match years later. Both were decided in three sets, although Perry's win over Jack Crawford was a more one-sided affair.

Both men delivered the long overdue Wimbledon title to great acclaim. Yet both Perry and Murray (to a degree) were outsiders. Perry was a working-class northerner who

didn't fit the Wimbledon mould. He was ahead of his time in several ways: he wasn't above using mind games to get an advantage, and had his own racquet maker and specially designed shoes. Perry was treated shabbily by the tennis elite, and was snubbed in the dressing room after his victory, with an All England official telling Crawford that 'the best man lost' and giving the runner-up a bottle of Champagne.

Murray, too, initially had a bumpy ride. A proud Scot at the All England club, in the early part of his career he played down the pressure at Wimbledon, and said he preferred the atmosphere at the US Open. He duly received a mixed reception from the crowd and media as a result. Time and success have made for a warmer relationship. Murray has not received the snobby treatment from the club that Perry did. He too professed to be 'disgusted' in the locker room post-victory, but merely by the taste of the champagne that was being sprayed about – Murray is teetotal.

Perry to Murray – the triumphant outsiders, separated by a lifetime. Which got *Sports Geek* thinking: can you link Perry to Murray via matches played only at Wimbledon? Could you do it just through Wimbledon champions? Could you even do it just through finals?

It's a tricky task: there are 39 different winners in the 78 years between Murray and Perry. There's a big gap for the Second World War when the tournament was suspended, and the problem of professionalism, when players were ineligible to play at Wimbledon until the Open era began in 1968.

Surprisingly, it turns out that you can get from Murray to Perry in just 17 steps, eight of which are finals, with nine in other rounds, all with past champions.

The chart shows those matches plus a few others that were notable, but aren't needed to complete the chain from Murray to Perry – it seems a shame not to have included Sampras vs. Agassi, or Ashe vs. Connors.

What emerge along the way are clusters where great rivalries have played out. The current golden crop of Murray,

From Murray to Perry

A generational history of Wimbledon champions

Wimbledon final
Other Wimbledon match

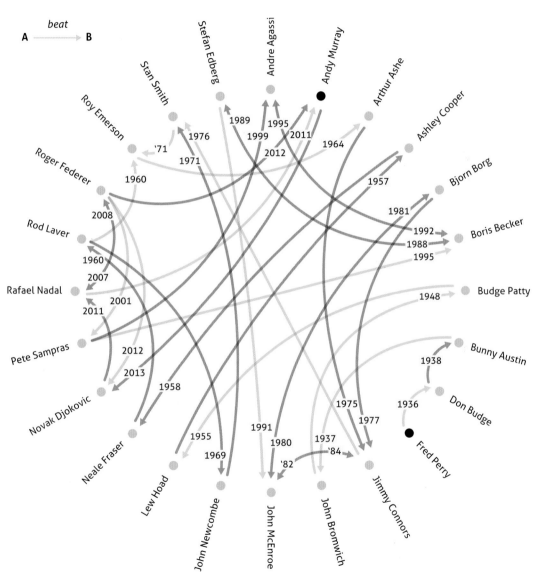

beat

A ———→ B

Stefan Edberg
Andre Agassi
Andy Murray
Arthur Ashe
Stan Smith
Ashley Cooper
Roy Emerson
Bjorn Borg
Roger Federer
Boris Becker
Rod Laver
Budge Patty
Rafael Nadal
Bunny Austin
Pete Sampras
Don Budge
Novak Djokovic
Fred Perry
Neale Fraser
Jimmy Connors
Lew Hoad
John Bromwich
John Newcombe
John McEnroe

1989 1995 2011
1999
'71 1976 2012 1964
1971 1957
1960 1981
2008 1992
1988
1960 1995
2007
2001 1948
2011
2012 1938
2013 1936
1958 1975 1977
1991 1937
1955 1980 '84
1969 '82

Djokovic, Rafael Nadal and Roger Federer have all faced each other at Wimbledon. Another cluster that stands out is the Borg-McEnroe-Connors era of the late 1970s and early 80s.

Occasionally – and surprisingly – certain tennis eras hardly overlap: some of Wimbledon's great champions never played each other on the grass of SW19. There is no McEnroe vs. Becker, or Sampras vs. Edberg.

As you go further back, it gets harder and harder to link players – but it can be done, just.

The tennis paradox: win the battle, lose the war

Winning the most points can still mean losing the match. What does it tell us about the top players?

'God, it's killing me.' As the tears welled up, those four words were all Roger Federer could say at the end of the 2009 Australian Open. This was the raw pain of defeat laid bare in front of millions. His great rival Rafael Nadal had just beaten him in another major final, having won in Paris and Wimbledon the year before. It was simply too much.

What Federer didn't know at the time was that he was actually the winner that day. In the match he won 174 points to Nadal's 173. He just didn't win the points that mattered most.

The scoring system of tennis is a thing of beauty. A game counts the same whether it goes to deuce, or is won to love. A 7–6 set counts the same as a 6–0 set. So a player can win fewer points but still win the match. We will call these paradox games: they happen around 5 per cent of the time on the men's tour and in Grand Slams. The cliché about winning the big points is true.

Frustratingly, data available from the women's tour, the WTA, doesn't give enough history on total points won in every match for a detailed analysis. The men's ATP tour, however, has the history of the breakdown of points won and lost for each player in each match for a quarter of a century.

Running the numbers for 1991–2015, it's hard to draw many conclusions. Some players that win more of these paradox games than they lose may be big servers, like John Isner, or go-for-broke hitters like Fernando Gonzalez. Quite a few are predominantly clay court players, like Carlos Moya and Tommy Robredo.

This may be because big servers often conserve energy when returning serve in order to maximise their own service games, especially if they are a break up in a set.

Clay court players end up in a lot of paradox games as clay matches sometimes become temporarily one-sided, mainly because it is a slower surface that makes the serve less dominant. This is illustrated by the higher number of 6–0 sets on clay.

The big surprise in the numbers is the player on the chart in the bottom right corner, with a high percentage of career wins (over 80 per cent), but a very low percentage of paradox games (under 25 per cent): Roger Federer.

The 2013 paper by Benjamin Wright, Ryan Rodenberg and Jeff Sackmann in the *International Journal of Performance Analysis in Sport*, which analysed paradox matches up to 2011, suggested that Federer's outlier position in this regard does not undermine his reputation. They suggested that his abnormally low percentage of paradox matches might be because during his peak, his opponents played in a high-risk style as they felt it was the only way to beat him.

They also noted that:

The results also imply that Federer does not engage in any short-term strategic tanking while playing. Even in matches he ultimately loses, the losses are rarely lopsided and very few games are non-competitive ... In sum, Federer's poor record is, in a somewhat ironic way, yet another piece of statistical evidence pointing towards the reasoned conclusion that Federer is one of the greatest tennis players ever.

Federer's poor record in paradox games is, paradoxically itself, proof of his greatness.

Making a point

Win percentage vs. paradox win percentage

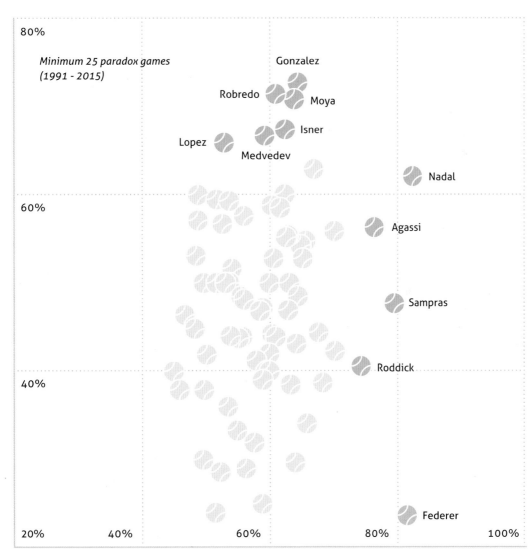

Minimum 25 paradox games (1991 - 2015)

Gonzalez

Robredo — Moya

Lopez — Isner

Medvedev

Nadal

Agassi

Sampras

Roddick

Federer

Paradox match win percentage (y-axis): 80%, 60%, 40%

Career win percentage (x-axis): 20%, 40%, 60%, 80%, 100%

Women tennis players are still losing the battle for equal pay

The gap has narrowed, but only for the few

Who earns more, male or female tennis players?

Percentage difference in male vs. female earnings (2001-2015)

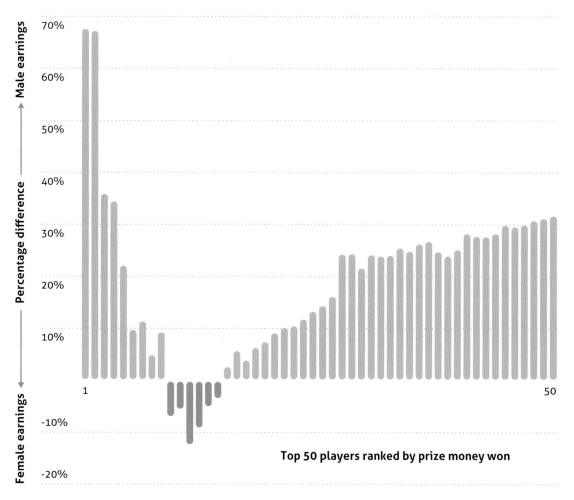

Top 50 players ranked by prize money won

When Billie Jean King beat Bobby Riggs in the 'Battle of the Sexes' in 1973, it was a triumph for women's tennis. Yet it took until 2007 for all four Grand Slam events to award men and women equal prize money.

Leave aside the question of whether it is right for women to be paid the same amount as men. There are reasonable arguments on both sides, which have been discussed elsewhere in great depth. The mistake is to think that because of the equal prize money at the Grand Slam events, men and women get paid the same for playing tennis. They don't.

The Slams are only one part of the picture. There's the rest of the year, when tennis players are at the smaller events, and this is where the disparity becomes clear.

In 2015, outside of the Majors, the average prize money on offer at a tour event on the men's ATP tour was $1,273,000. For women on the WTA tour, it was $989,000 – a difference of nearly 30 per cent.

The difference in total tour prize money is even starker. As the men's tour features more events, the total ATP prize money pot is just under $79m, compared to $56m for the WTA: that's 40 per cent higher.

And it gets worse. Tennis tournaments publish their Total Financial Contribution (TFC), which includes the prize money, various fees, and also the contribution to the bonus pool. The bonus pool goes to the players at the end of the season for turning up to the bigger events.

Comparing the TFC of the two tours, the average men's event goes up to $1,482,000 per event, compared to the women's $1,108,000. The gap has increased to nearly 34 per cent. In terms of total TFC, the men's $92m is 46 per cent higher than the women's $63m.

Strangely though, you would be better off being the 10th best woman in the world than the 10th best man. For the last five years (2011 to 2015), the prize money won by the top men is higher for the top few, but then drops off very rapidly. The women's prize money chart is a less steep curve.

However, once you go past the top 20, the men earn around 20 to 40 per cent more than the women. Each year shows the same pattern, although the chart shows the average over the five years.

Some years are more bizarre than others. In 2011, nine of the top 14 women in the world earned more than their male equivalents.

The highest rank where a woman out-earned a man was number 3 in 2012, when Maria Sharapova took home nearly $1m more than Andy Murray.

The top 10 to 15 on the men's tour are earning less than the women because the top few men, mainly Roger Federer, Rafael Nadal, Novak Djokovic and Murray, are almost exclusively scooping the big prizes. It's slim pickings for the rest.

Giant killing in tennis

Which is more likely to see a big upset: a men's or women's match?

Rafael Nadal might want to get hold of footballer Mario Balotelli's T-shirt that says 'Why always me?' Nadal's recent Wimbledon results are taking on a familiar pattern: scrape through one round or two, and then get beaten by a player ranked outside the top 100. Players such as Dustin Brown, Lucus Rosol, Steve Darcis and Nick Kyrgios.

But should we be surprised? Are tennis upsets on the increase? Given the improvements in training and competition, you might think that lower ranked players are beating those higher up the ladder with increasing frequency. We are always told that 'on the day', any player can beat one of the top guys. The women's game, on the other hand, apparently has less strength in depth: the top players aren't challenged until later rounds in a tournament.

Think again. The reality is that big upsets – where a player ranked outside the top 100 beats a player in the top 10 – are on the decline in the men's game, and on the up in the women's. (Matches include all grand slams and the ATP and WTA tours.)

If we look at the number of upsets per 1,000 matches since 2007, the men's rate has dropped from over 6 to 1.5. The big upset is four times as rare as it once was. For the women, the opposite is true: from less than one match in a thousand, there are now nearly six.

Why is this happening? This is partly because upsets are rare anyway, so a few matches makes a big difference. Year to year, you can count the change in the number of big upsets on two hands.

In the men's game, the top 10 have been very stable for the past decade, and players such as Roger Federer, Novak Djokovic and Andy Murray rarely crash to humiliating defeats. On the women's side, the top ten have been less dominant, Serena Williams aside.

Nadal seems a special case. He leads the way in current players to suffer big defeats, with ten losses to 100-plus ranked players (while ranked 10 or higher). His rivals, Federer, Djokovic and Murray, have ten such defeats combined.

What is it about Nadal that makes him prone to such upsets? Perhaps it's his style of play. Or perhaps players fancy their chances against him away from his favoured surface of clay – all of his big upsets have come on grass and hard courts.

Don't get upset

Number of upsets, women vs. men

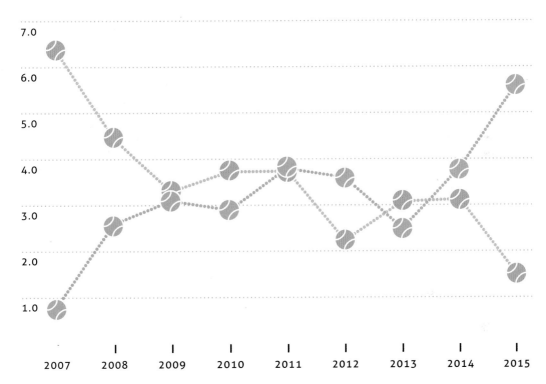

Parkrun

The biggest club you've never heard of

Is parkrun the world's largest participatory event? All movements and revolutions have a starting point. For parkrun, it is Bushy Park in London, on 2 October 2004. Organised by Paul Sinton-Hewitt, the 13 people who took part in what was then called the Bushy Park Time Trial had little idea that they would be the first of over 1 million runners worldwide.

Parkrun is a simple idea. The guiding principles are: 'weekly, free, 5km, for everyone, forever'.

It took a while to take off – two years went by before the second event in Wimbledon. It wasn't called parkrun until 2010. But as the chart shows, it has exploded in terms of the number of weekly runners. There are now parkrun events across 11 countries in over 700 locations. The total number of runners is now over 1.2m, relying on the goodwill of 150,000 volunteers who set up the courses and register the times. Over 100,000 people take part each weekend.

It is still primarily a UK event, with Bushy Park the spiritual home – the parkrun there often attracts over 1,000 runners, with the record being 1,705 on the 10-year anniversary run. Although an amateur event, the parkrun record is quick: at 13m 48s, it would be good enough to provisionally qualify for the US Olympic team.

How big is parkrun? The best comparison we can find would be with health clubs: although parkrun is free and not a commercial enterprise, like health clubs it requires people to motivate themselves to take part.

If we count all parkrun runners as members of one club, it is certainly big, but not the world's biggest – in the US, Planet Fitness has 4.8m members. However, it is now comparable in size to Virgin Active, which has 1.4m members worldwide. For a community-run, volunteer-reliant system, that's no mean feat. And given the current growth rate, it will not be long before it may legitimately have a claim to be the world's biggest sports club.

12	1,874	194,000	107,000	195
Countries	Highest attendance, at North Beach, New Zealand, June 18 2016	Volunteers worldwide	Events held worldwide	Runners that have completed over 100 runs

920	1.6m	13m 48s	15m 43s	16m
Total number of park locations	Total runners worldwide	Male parkrun record	Female parkrun record	Total runs completed

Racking up the miles

Avg. Run Time	Events	Runners	Locations
Australia 0:31:08	15,950	226,196	167
Denmark 0:25:34	2,014	8,993	7
Ireland 0:28:40	3,391	64,935	52
New Zealand 0:29:22	1,261	14,322	11
Poland 0:25:29	3,014	23,399	35
South Africa 0:39:51	7,042	243,401	79
UK 0:27:40	63,540	976,098	401
USA 0:30:41	561	6,320	5

Is the Tour de France getting easier?

If you look at the distance and number of finishers, perhaps it is

The Tour de France is one of the most gruelling events in the world. Over 23 days, the best cyclists battle through huge distances, mountain climbs and sprint time trials. Many amateur cyclists see it as a great challenge just to complete one stage, let alone 21 stages back-to-back.

If you tried to tell one of the riders that the Tour de France is getting easier over time, you'd get pretty short shrift.

But one of the great things about analysing data is that it can show things that we can't otherwise see. Our gut instinct is that sport gets tougher and more competitive every year – including the Tour de France.

The numbers tell a different story.

Leave aside for a moment the question of training, the teams and support that the riders now get. These are hard to measure, but clearly over the last century have improved beyond recognition. Let's also leave aside for now the question of whether the riders have got quicker, allowing for the improvement in bike technology. Let's also leave drugs out of it.

Instead, let's look at one very basic question: how many riders finish the race? This is a simple measure. Riders 100 years ago would have ridden as hard as they could to win, just as they do now.

Over time, more and more riders have finished the course. The number of finishers hit 80 per cent in 1974 for the first time, then again in 1981, and has since been over 80 per cent nine more times in the last 14 years.

What makes riders drop out? Well, it's not just physical exhaustion. The big dip in the finishing rate in 1998, when only 51 per cent of riders completed the Tour (sandwiched between 70 per cent in 1997 and 78 per cent in 1999) was due to a massive doping bust, with riders being arrested and teams pulling out.

Another big reason why riders pull out is if they crash. Some years have seen big pile-ups in the main pack – the peloton, which can cause injuries or push riders so far behind that they see little point in carrying on.

There are other reasons that riders don't finish. Sprint specialists have been known to pull out of the Tour for the mountain stages, with Italian Mario Cipollini famously boasting about being on the beach while others were slogging it out on the hills. It didn't go down very well with other Tour riders.

So why are finish rates going up?

Cyclists are still being caught doping – that much hasn't changed. Granted, 1998 was an exceptional year, but 2007 and 2013 were big years too in terms of drug scandals, and the finish rate is still historically much higher than in the pre-drug testing era.

It's also true that some of the sprint specialists are competing in the full Tour, rather than dropping out. However, these are a small number of cases, and won't change the overall rate that much.

So let's look at the composition of the Tour. One thing to note is that the number of stages has fallen. From 1927 to 1987, it was frequently as high as 24 stages, and most often 22. Since 1992, it has only been 20 or 21. While this doesn't sound like the biggest change, an extra stage or two can break a cyclist – especially when the organisers made riders do two stages in one day – and only adds to the chances of a crash.

Combined with the fewer stages is the shorter distance. From the peak of 5,745km in 1926, the race distance has fallen steadily to around 3,500km. And that isn't just through having fewer stages – the average distance per stage has fallen too, from around 300km per stage to closer to 150km (see chart). Of course, the tour changes the stages each year, and has added some gruelling climbs to the race. But there have always been tough stages in the Alps and Pyrenees, as well as the faster, flatter stages.

Overall, it's a shorter race, with fewer stages. Does that mean it's easier? It's worth asking whether the drop in race distance per stage and corresponding increase in finishing rate are simply a coincidence, even if it might not be a popular theory.

Tour De France: shorter and easier?

● Finish rate ● Distance per stage

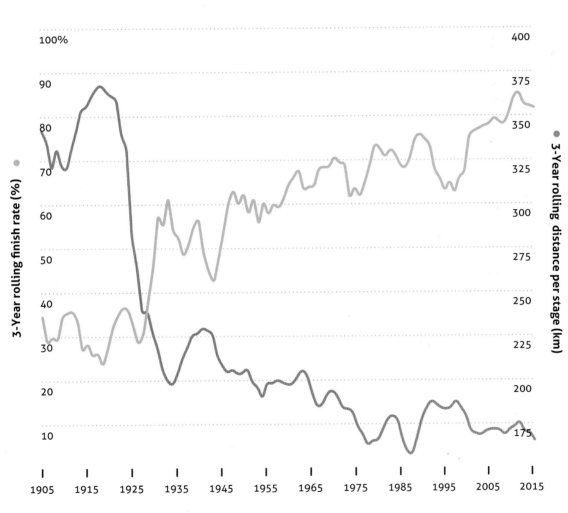

3-Year rolling finish rate (%) ●

● **3-Year rolling distance per stage (km)**

Burden of proof

Cycling doesn't have a drug problem: the clean riders do

When a cup of urine was thrown in Chris Froome's face during his victorious 2015 Tour de France, it wasn't simply written off as the act of a deranged spectator. Froome's team, Sky, were quick to blame parts of the French media that had looked to discredit their rider's performances, implying that he was on drugs.

What could Froome do? In a sport that has had so many high-profile doping cheats, any riders who win well are automatically deemed under suspicion.

The irony is that cycling itself is doing fine, despite the scandals. Crowds still line the streets, cycling at the Olympics as well as the main tours has healthy TV coverage. Teams are still investing in the sport. Considering the scandals, it's remarkable that it hasn't completely fallen apart.

Yet the clean riders bear the brunt. As the saying goes, absence of evidence is not evidence of absence. How can you prove you are clean? The chart shows how cycling violations, positive tests and admissions of doping have been coming thick and fast for years.

The spike in 1998 was partly due to the Festina team at the Tour de France, when all nine riders confessed to EPO usage after drugs were found in a team car. The increase in violations around this time was also partly due to more sophisticated blood testing, which looked for abnormally high haematocrit levels, or numbers of red blood cells by blood volume.

A history of shame: cycling doping incidents

● Admission statement ● Positive test ● Violation

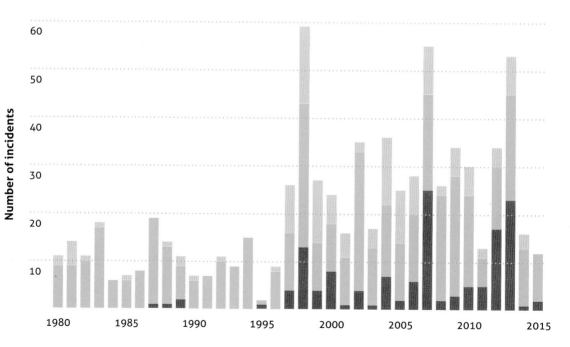

The systematic and widespread use of drugs in cycling is shocking, but it has always been the way. Back in the 1930s, the Tour de France rulebook acknowledged the usage when it made it clear that the race organisers would not provide drugs. Great riders have always been caught, or confessed: from Eddy Merckx in the 1970s to Marco Pentane and Jan Ullrich all the way to the biggest catch of the lot, Lance Armstrong.

In effect, the chart shows the cat-and-mouse game of better testing against better disguise. A look at the drugs involved shows just how crucial detecting erythropoietin, known as EPO, has been. EPO stimulates red blood cell production, so an athlete can carry more oxygen. It also has tragic side effects, mainly that the thicker blood can lead to heart attacks. We will never know for sure how many cyclists have died from taking EPO and other similar drugs.

Nor will some people ever be satisfied that some riders are clean, even if they go to the extraordinary lengths to publish their performance data and history, as Froome has done.

What's your tipple?

Number of times a product was involved in a doping incident (1980 onwards)

343	Erythropoietin (EPO)
341	Undisclosed
125	Testosterone
122	Human Growth Hormone (hGH)
107	Blood transfusion
67	Corticosteroid
55	Amphetamine
50	Cortisone
47	Nandrolone
36	Continuous Erythropoietin Receptor Activator (CERA)
34	Caffeine
34	Contractual violation
29	Darbepoetin (NESP)
28	Bio passport irregularities
28	Missed test
26	Cocaine
25	Clenbuterol
25	Unspecified steroids
21	Salbutamol
20	Human chorionic gonadotropin (hCG)
19	Actovegin
19	Insulin
14	Ephedrine
13	Heroin

The hour mark: back from the dead

Cycling's most prestigious record has a confusing history but a bright future

There's something wonderfully pure about the one-hour cycling record. No tactics, other than maintaining your pace. No other riders to get in your way. Just rider, bike, and a clock. No wonder it is probably the most prestigious cycling record.

Cyclists describe it as the toughest thing they have ever done. Jack Bonbridge said after his (failed) attempt on the record in 2015 that it was 'the closest [he] could feel to death without actually dying'. Others have used similar words. Sir Bradley Wiggins described it as torture, and compared it to giving birth.

The other thing about the hour record is that it has to be attempted. You don't just happen to set it as part of another race. It is organised, publicised, scrutinised. You have to be brave to put yourself forward in such circumstances.

Yet for all the prestige, the record is a historical mess — mainly thanks to bike technology. Let's try to make sense of it.

A quick recap: the progression of the hour record came to a crossroads in early 1984. The mark set by the great Eddy Merckx back in 1972 was 49.431km, and stood for 12 years. It was beaten by Italy's Francesco Moser, pushing it to 50.808km, and again by Moser over the 51km mark four days later.

Moser had help: specifically, disc wheels and a special suit. His record stood for nine years, until in 1993, helped by new technology and controversial racing positions, Graeme Obree and Chris Boardman pushed it over 52km, followed by Miguel Indurain and Tony Rominger. In all, the record was set six times in just under 16 months, going from 51.151km to 55.291km — an improvement of over 8 per cent on the Moser record. Finally, in 1996, Boardman set a new mark of 56.375km.

Then it all went quiet. The *Union Cycliste Internationale* (UCI) decreed that all the records set since 1984 were aided too much by technology, and so paradoxically called them 'best human effort'. The old UCI record of Eddy Merckx still stood.

That was until 2000, when Boardman, again, took the record, but this time on a traditional bike. He cycled 49,441m – just 10m further than Merckx had done in 1972. Until 2014, there was just one more record, the UCI-approved mark of Ondřej Sosenka in 2005, a rider since found to be doping.

For years, the hour record stood unchallenged, mothballed.

Then in 2014, the UCI clarified the rules on which bikes could be used. It sparked a flurry of records. First Jens Voigt set the record at 51.110km in September 2014, then Matthias Brändle pipped that by 742m one month later. Rohan Dennis pushed it over 52km in February 2015. In May, Alex Dowsett set a new record, just 3m shy of 53km. And then Sir Bradley Wiggins pushed on to 54.526km in June. Since going to press, there may have been another series of attempts.

Overall, five records were set in less than 9 months – even quicker than the 1993–94 period, and with a similar increase: 9.7 per cent from the old Sosenka record. This time though, all on approved bikes.

This period of tit-for-tat record setting isn't unusual. If we look at the hour record back over the last 100 years, it has often been set in small clusters, with the record quickly changing hands before a period of calm.

This recent surge, however, is unprecedented. Other periods have seen increases of 3 to 6 per cent over a couple of years – this is nearly 10 per cent in less than one year.

So we are in new territory, both in the rate and margin of increase in the hour mark. Given that the record had a long hiatus, combined with modern training and the current crop of talented riders, perhaps we shouldn't be surprised. The question is where the upper limit will be found. Is it 55km? 56?

Wiggins' record may stand for over a decade, like some before. Or it may be bettered much sooner. Either would fit perfectly well with the patterns of the past.

History of an hour

● UCI record   ● Best human record

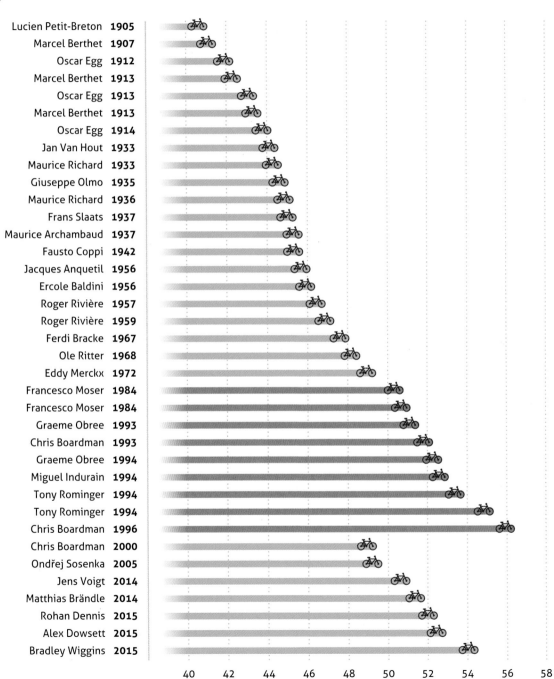

Lucien Petit-Breton	**1905**
Marcel Berthet	**1907**
Oscar Egg	**1912**
Marcel Berthet	**1913**
Oscar Egg	**1913**
Marcel Berthet	**1913**
Oscar Egg	**1914**
Jan Van Hout	**1933**
Maurice Richard	**1933**
Giuseppe Olmo	**1935**
Maurice Richard	**1936**
Frans Slaats	**1937**
Maurice Archambaud	**1937**
Fausto Coppi	**1942**
Jacques Anquetil	**1956**
Ercole Baldini	**1956**
Roger Rivière	**1957**
Roger Rivière	**1959**
Ferdi Bracke	**1967**
Ole Ritter	**1968**
Eddy Merckx	**1972**
Francesco Moser	**1984**
Francesco Moser	**1984**
Graeme Obree	**1993**
Chris Boardman	**1993**
Graeme Obree	**1994**
Miguel Indurain	**1994**
Tony Rominger	**1994**
Tony Rominger	**1994**
Chris Boardman	**1996**
Chris Boardman	**2000**
Ondřej Sosenka	**2005**
Jens Voigt	**2014**
Matthias Brändle	**2014**
Rohan Dennis	**2015**
Alex Dowsett	**2015**
Bradley Wiggins	**2015**

40 42 44 46 48 50 52 54 56 58

Record distance in an hour (km)

In defence of Sally Robbins

The rower that gave up didn't cost her team a medal

Rowing is in some ways the ultimate team sport. Unlike in football, where moments of individual brilliance can turn a game, in rowing you are only as strong as your weakest teammate. The whole crew must perform – anything less, and the opposition will beat you.

Giving up is akin to sacrilege in rowing. Type 'rowing controversy' into Google, and the Australian Sally Robbins still comes up as the top result. The story is in essence quite simple. In the last part of the final of the Women's Eight at

the 2004 Athens Olympic Games, she gave up, completely exhausted. She simply stopped rowing. She even lay down at one point. Australia finished last, more than 13 seconds behind the winners, Romania.

The aftermath was brutal. Robbins was deemed 'almost un-Australian' by Cathy Freeman, the 400m champion, and was dubbed 'Lay Down Sally' by the media. Her fellow crew members, rather than rally round her, criticised her publicly. She was hit by another rower at an athletes'

Don't stop now

The women's eight 2004 Olympic final

● Romania ◍ USA ● Netherlands ◌ China ● Germany ◍ Australia

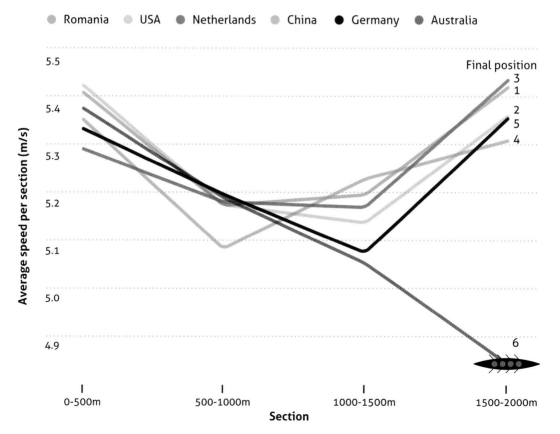

dinner. 'It's eight, mate, pull your weight' was one headline in Australia.

Did she cost the team a medal, as has been asserted by the *Telegraph* and several other publications and in a book (*Don't Rock the Boat* by Peter Wilkins, 2008)?

The basic facts would suggest so. A rowing race is 2000m. According to most reports, at 1500m, Australia were in third place. Then Robbins stopped rowing and lay down with either 400m or 250m to go (reports differ on the exact moment). The only TV footage available is unclear. The BBC commentators at the time make no reference to Robbins giving up until very close to the finish line.

So what's the case for the defence?

The Australian crew were not expected to win this event, as has also been pointed out in several publications. They had scraped into the final, coming fourth in their heat and then third in the repechage (the last chance qualifying round). The gold was not a realistic target. But was bronze, or even silver a possibility?

The chart shows the split times that each crew registered over the four 500m sections of the race. The last block shows the Australians way back, with a 1500m to 2000m time of 103 seconds.

But dissecting the race, a different story emerges. Australia had, in fact, already blown it by 1500m. Their time for the third part of the race, from 1000m to 1500m, was a fraction under 99 seconds, over two seconds slower than the Netherlands crew who finished third. It was up to that point the slowest 500m split of the race.

Far from being in third place at 1500m as was stated in several news reports, Australia were actually fifth.

Of course, we could point the finger at Robbins for this – perhaps she was already slowing the team down. But given that the reports of the incident state she collapsed after the 1500m mark, we have to collectively blame the Australians at this point.

So had Robbins rowed on, could Australia have got bronze? To do that, they would have had to row significantly faster than the third-placed Netherlands in the final 500m, who put in a storming finish. In fact, they would have had to row the final 500m in 91.42 seconds to pip the Netherlands to bronze even by 1/100th of a second.

Let's put it in context. That would have meant rowing *half a second* faster than any crew had rowed 500m of the race, having just posted the slowest. (The quickest 500m was 92.03, by the Netherlands in the final section). Even the winning Romanian crew never went below 92 seconds.

In other words, before Robbins stopped, the Australians' race was over. There was simply no way they were going to win a medal. That final 500m was not possible, even with all eight rowers performing at their best.

Of course, this will miss the point for many Australians. Robbins gave up, and you just don't do that. But to blame her for the whole team missing out on a medal is unjust, and ignores the facts of the race. If we assume Robbins was rowing to the best of her ability in the first 1500m, the Australian crew weren't going to get a medal, whatever Robbins did next.

That's the price you pay

Which is more expensive, the World Cup or the Olympics?

Build several new stadiums, upgrade a rail network, add in a new airport, perhaps. A media centre, a new highway, and several new hotels. Hosting an Olympics or a World Cup is, in effect, a series of controversial and expensive construction projects with a few weeks of sport tacked on at the end.

Which is more expensive? At first glance, it should be the Olympics: the range of sports facilities needed presents

a huge challenge. Each summer games needs a top-class velodrome, aquatics centre, tennis complex, kayaking course, basketball arena, and so on. Most cities in the world that could consider an Olympic bid might have a few places ready to use, but not the full list.

A football World Cup, by contrast, just needs some decent football stadiums. Any country with a reasonable

The price you pay

The initial bid vs. final costs of major sporting events and cost per spectator

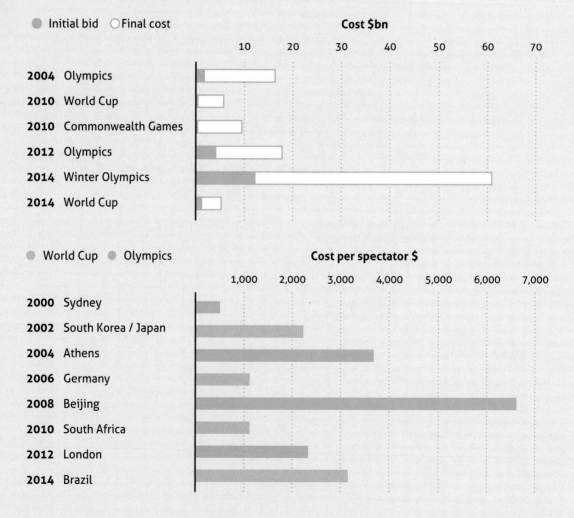

league will have a handful of grounds that meet the need. Add in an existing national stadium for finals and host matches, plus temporary expansion of a few others, perhaps 'borrowing' a stadium from another sport, and you're ready to put on the tournament, surely?

Not quite. Although Fifa claims not to make bid stipulations, leaked documents show that it makes heavy demands of host countries in terms of building stadiums.

According to a 2012 report by the Play the Game/Danish Institute for Sports Studies, Fifa requires that the main venue used for the opening game and final must have a capacity of at least 80,000. Semi-finals must be able to seat 60,000 spectators, and other matches need stadiums with 40,000 capacity. In all, there should be 12 venues.

This can lead to poor choices; stadiums are sometimes built in parts of a country that don't have a local team to sustain a stadium of that size. It happened in South Africa and Brazil and will almost certainly happen in Russia in 2018 and Qatar in 2022. After the event stadiums often lie derelict, sometimes used as car parks. Yet the construction continues.

To compare the costs of a World Cup and a Summer Olympics, *Sports Geek* looked at the total cost of the event, and the official attendance or tickets sold. Using those two figures, we can get a total cost per spectator. This is not what the spectator pays, but what it cost to stage the event.

Where possible, the official costs announced by each country have been used. Some figures are tricky: the Beijing Olympics is based on the most reliable cost estimate by various economists. It's hard to know if various infrastructure costs and other facilities that aren't strictly about the event were put into the overall figure. So overall, the figures come with a (mild) health warning.

What we find is that while a World Cup in theory should always be cheaper than an Olympics, it's not. In terms of the total cost per spectator, the bill is sometimes much higher. For example, Brazil 2014 cost around $3,000 per spectator, while the London Olympics in 2012 was $2,300. Beijing was a hugely expensive Olympics, blowing over $6,600 per spectator, while the 2010 and 2006 World Cups were much cheaper affairs, at around $1,000 per head. In contrast, the 2002 World Cup (jointly hosted by Japan and South Korea, and consequently using more stadiums than normal) cost nearly as much as London 2012, and four times as much as the preceding Olympic games in Sydney.

According to Play the Game, when assessing the cost of the main stadiums constructed since 2000, the average price per seat for the Olympics was $4,176, compared to $4,525 for the World Cup.

Will this change? It depends on the governing bodies and the bidding process.

The International Olympic Committee passed new rules in 2014 to encourage a more affordable bidding process and hosting, allowing for Olympics to stage events outside the host city or country in order to keep the costs down, as well as other wide-ranging reforms. This was partly in response to the embarrassment of the 2022 Winter Games bidding process, in which many cities pulled out, leaving just Beijing and Almaty of Kazakhstan. Beijing, a city with no snow, won.

Fifa has (at the time of writing) not reformed the bidding process; the corruption controversy and arrests that engulfed the organisation in 2015 have left it in disarray.

However, for some World Cup and Olympic hosts, the weakened state of their economies means that the spiralling costs for any large event are no longer an option. Russia, which blew billions on the Sochi Winter Olympics of 2014, has cut its budget for the 2018 World Cup, scaling back on spending on stadiums, hotels and infrastructure. Japan has also trimmed the costs of the 2020 Tokyo Olympics, shelving an ambitious main stadium plan for something more modest. Even rich Qatar, which controversially was awarded the 2022 World Cup, has had to cut back, building eight stadiums rather than the originally planned twelve.

One day bids may be evaluated on how inexpensive they are and how facilities can be adapted post-event. Until then, taxpayers will be saddled with exorbitant stadiums that look great on TV, but post-event just gather dust.

Kangaroos rarely caught

The best thing that could happen to Rugby League would be Australia to lose more often

Sport needs uncertainty. Otherwise, what's the point?

For as long as many fans could remember, if and when the Rugby League World Cup came around, Australia won it.

The tournament itself has been blighted by irregularity: the four-year cycle of other world cups has rarely happened. The interval in years for the Rugby League World Cup since the inaugural event of 1954 has been:

3, 3, 8, 2, 2, 3, 2, 11, 4, 3, 5, 8, 5, 4 (to 2017)

The number of teams taking part has been:

4, 4, 4, 4, 4, 4, 5, 4, 5, 5, 10, 16, 10, 14, 14

Those sequences of numbers tell their own story. The Rugby League World Cup has been one of disappointment and chaos. The trophy even went missing for 20 years, from 1970 to 1990. Against this backdrop of organisational chaos the sport has also suffered from a high degree of predictability.

International Rugby League for the past 30 years has been dominated by Australia. The chart shows the 10-year rolling win rate for the three big teams – Australia, along with New Zealand, and England/Great Britain (treated as one for the purposes of this analysis).

Australia surges ahead in the 1980s, never dropping below an 80 per cent win rate from 1982 onwards – a quite phenomenal record.

Until 2008. Australia went into the World Cup as clear favourites as usual, having won the previous six events. But New Zealand pulled off a shock victory in the final, winning 34–20. It was a result that defied not just history, but all predictions.

Normality was restored in 2013, with Australia winning their last three matches including the final by an aggregate score of 160–2. At least the 2008 New Zealand victory showed that the World Cup could be unpredictable, even if 2013 went as expected. It's also good that the next two World Cups are scheduled in 2017 and 2021: a normal four-year cycle.

Australia ascendant

The 10-year rolling win rate for the top three Rugby League teams

● Australia ● New Zealand ● Great Britain / England ★ World Cup winner

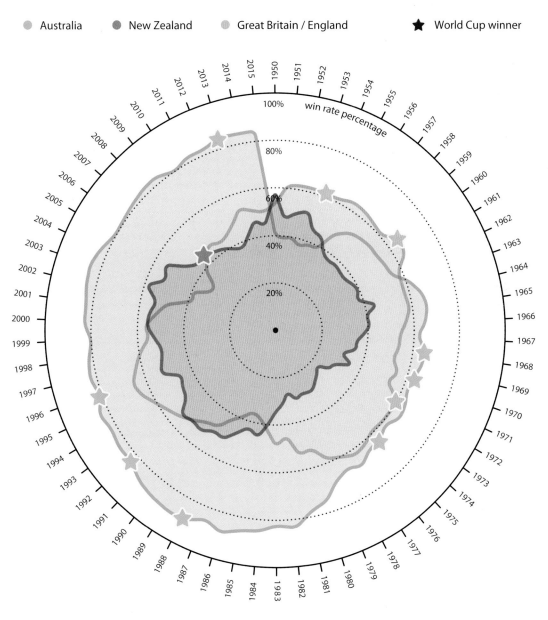

win rate percentage

100%

80%

60%

40%

20%

The not-so-Super League

It was the original professional breakaway sport – but Rugby League has failed

What do you do when your rival starts wearing your clothes, even if they are 100 years old?

For Rugby League, when the Union code of the game followed suit and turned professional in 1995, the answer was to become a summer sport. Other innovations – such as abolishing relegation – have hampered the sport.

The origins of League date back to 1895, when rugby clubs in the north of England split to form their own union, mainly over the issue of paying players.

From then on League dominated in the north of England, while Union was a southern sport, with one or two club exceptions. Union remained a 15-a-side game, with line-outs and scrums. League was a 13-player version, with no real forward set-pieces and more running.

In the amateur era, traffic was one way: players crossed to the League code, leaving Union for the money. But then, once Union turned pro in 1995, the flow was reversed. Players with the ball-handling skills and speed honed in League took up Union, such as Jason Robinson. Some, like Jonathan Davies, came back to their roots.

So, to create some difference, in 1996, League became a summer sport. And for a while it worked – attendances went up.

A look at the average crowd from both the Super League (League) and the Premiership (Union) shows how League has fared since. Crowds have peaked, and are gradually falling, while the union-code Premiership surges on.

It doesn't help that League is out-marketed by Union, which has a far healthier international scene. League has nothing to rival the 6 Nations or Rugby Championship. And although there are 18 full member test nations of the Rugby League International Federation (including Lebanon and Jamaica), the Rugby League World Cup has a fraction of the impact of the Union World Cup.

There is another factor in declining match attendances: relegation. Fans want something to cheer about, and their team something to fight for. In any sports league there is usually a battle for the top spot; there is the lull in mid-table; and there is a dog-fight to avoid the drop. The Super League had relegation, with the lowest team dropping into the division below – until 2009.

From 2009 onwards, the Super League changed to a three-year licence system, where teams were assessed for inclusion based on their stadium facilities, financial performance, playing strength and commercial appeal.

The result? More meaningless end-of-season matches, which lowers attendances. You can see that League crowds peaked in 2007, followed by leaner years.

Of course, there are many other reasons why attendances might fall in a sport. If a promoted team has a smaller stadium than a relegated one, overall attendance will drop. Other sports may take fans' attention, such as a thrilling cricket series or a football World Cup.

Regardless, sports that have dwindling attendances are in trouble. Half-empty stadiums have a poor atmosphere and make other fans question turning up next time, as well as looking bad on TV. The Super League needs to generate excitement. Luckily, League has seen sense: relegation and promotion have been reintroduced for the 2015 season, as well as cutting the number of teams to 12 from 14, and revamping the play-off format. But is it too late to bring back the crowds?

Code battle

Average attendance of Super League and Rugby Premiership (Union) games

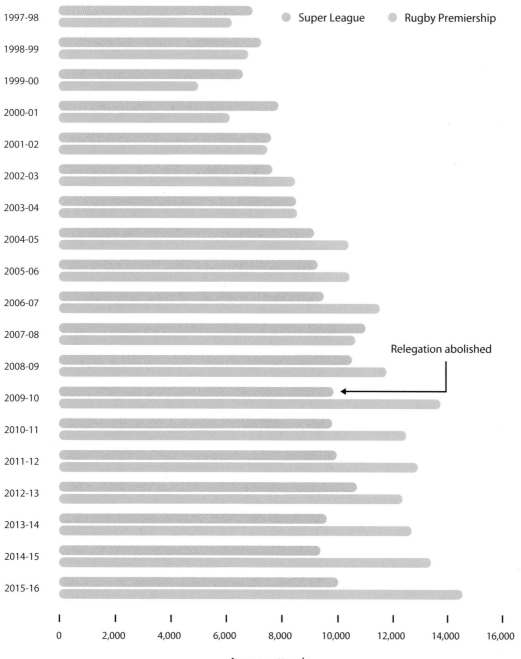

Michael Jordan is the greatest player ever

Or is he? LeBron James could be better

Some sports thrive on the constant rumbling debate about who is the greatest of all time. Pele or Maradona? Schumacher or Fangio? Williams or Graf? Other sports are blessed with a more clear-cut candidate. Bradman. Phelps. Jordan.

In most people's minds, Michael Jordan is the greatest basketball player of all time, without question. But if he is not being usurped, he is certainly being challenged by LeBron James. Here's how.

They are both statistically brilliant players; but is there one measure that could shed more light? One such measure is the Player Efficiency Rating (PER), invented by John Hollinger of ESPN to try and condense a player's game contributions into one statistic. As Hollinger put it: 'The PER sums up all a player's positive accomplishments, subtracts the negative accomplishments, and returns a per-minute rating of a player's performance.'

The PER is complicated, so we won't go through the formula here. What is interesting is that the ranking allows for cross-generational comparisons by setting up the league average at 15.0 points each season.

So who is top? Michael Jordan still wins out, with a PER of 27.9, just a fraction ahead of LeBron James in second place on 27.6. However, a couple of strong final seasons in his career, and James could move ahead. Would that be a bad thing?

It depends on your viewpoint, of course. For some, Jordan's scoring prowess, titles, athleticism and clutch play (scoring at the final moment to win the game) will always set him apart. James, though, is frequently regarded as a better all-round player.

The big mistake Jordan made was to come out of retirement and play for the Washington Wizards for two seasons, in 2001 to 2003, when he was approaching 40 years old. His PER scores for those seasons were 20.7 and 19.3 – still good numbers, but nothing like his years of PER scores over 30, and not in the top 10 in the league.

Had Jordan stopped and not come back in 2001, his career PER would have been 29.05 – quite a bit higher, and probably (if not certainly) out of reach for James.

Without those two extra years of decline, the debate would never even have surfaced. Sometimes it is worth staying retired.

Season by season

How Michael Jordan and LeBron James measure up by PER

● Michael Jordan   ● LeBron James

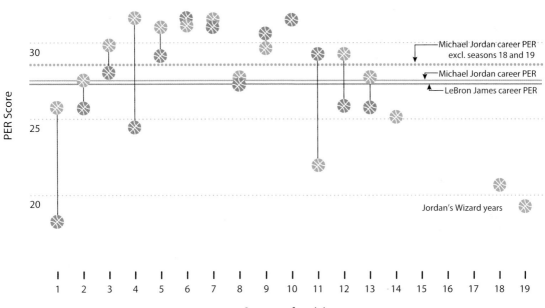

PER Score

35

30 ·········· Michael Jordan career PER excl. seasons 18 and 19

·········· Michael Jordan career PER

LeBron James career PER

25

20

Jordan's Wizard years

| | | | | | | | | | | | | | | | | | | |
|1|2|3|4|5|6|7|8|9|10|11|12|13|14|15|16|17|18|19|

Seasons after debut

Just a long-range shooting contest

Is the rise of the three-pointer bad for basketball?

The trade-off is a pretty straightforward one: score close to the basket and you get two points, a field goal. Shoot from behind the line further away, and it's three.

Simple as it sounds, the NBA is currently going through something of an identity crisis about the three-pointer.

It's easy to see why. The 3-point shot is on the rise. The number of attempts per game has risen from two for each team in 1980 to over 23 in 2015 – that works out at 47 3-point attempts per game for the spectators to sit through.

Many coaches and traditionalists are unhappy. Even teams that use the 3-pointer and win don't like it. Gregg Popovich, coach of the NBA championship-winning San Antonio Spurs in 2014 told ESPN: 'It mucks up the game.' The Spurs used the 3-pointer to devastating effect in the finals against the Miami Heat, averaging over 23 3-point shots per game.

That all sounds like a lot. But it needs context. Firstly, the NBA didn't always have the 3-pointer. It was introduced in the 1979–80 season, and took a few seasons to really take hold as a weapon. In the first few seasons, the success rate was less than 30 per cent of 3-point shots attempted.

Then the NBA tinkered with how close the line was to the basket, moving it closer around the central area to 22 feet for the 1994–95 season, from the original 23 feet 9 inches in front of the basket. This stayed until the 1997–98 season and then the line moved back, and the effect can be seen on the total number of 3-pointers per game – the chart has a hump in it.

Let's clear up the first myth: the changes to the 3-point line did not kick-start the change in behaviour of NBA teams. The rise in shooting from distance reverts straight back to the path it was on (before the closer line) from 1998 onwards.

But it *did* have an effect on 3-pointer accuracy. The closer line made players more likely to hit the three-pointer, but when the line was moved back, the accuracy remained. The difference in accuracy between field goals (2-pointers) and 3-pointers has not changed since the 1994-95 season.

That would explain the increase in use of 3-pointers up to 1995 – players were more accurate. However, the last 20 years has seen no significant change in the accuracy of 3-pointer attempts. So how do we explain the recent increase in 3-pointer attempts from 36 per game in 2010–11 to over 47 in 2015–16?

There is one key factor behind the rise of 3-pointers: it works. Teams that shoot a high number of 3-pointers, and are accurate, tend to win. And coaches have worked this out.

Before we write the NBA off as a glorified long-shot contest, as some commentators have, there is a case to be made that it's not that bad.

First, the recent rise in 3-pointer attempts is part of a rise in all attempts. Field goal attempts have gone up too, by around four per game in the last four seasons. That is fewer than the ten 3-pointers per game in the same period, but it is important. This is not a substitution problem: field goals aren't being shunned for three-pointers. The total number of shot attempts per game has gone up – from 190 in the 1990s to over 210 now.

This means that the overall composition of a game has changed, but not by as much as the rise in the 3-pointer would suggest. Looking at the percentage of attempted shots in a game that are either field goals or 3-pointers, the change in the last five seasons is noticeable, but hardly shocking. In 2010–11, 18 per cent of shots were 3-pointer attempts. In 2015–16, it was 22 per cent.

Which means that in a typical game, around four out of five shots (78 per cent) at the hoop are still what the traditionalists would like to see. That's not so bad. At the current rate of change, it will be another 50 years at least before the game is around half 3-pointers, half field goals.

Trying it out

Basketball scoring composition in the NBA

● Field goal   ● Three-point

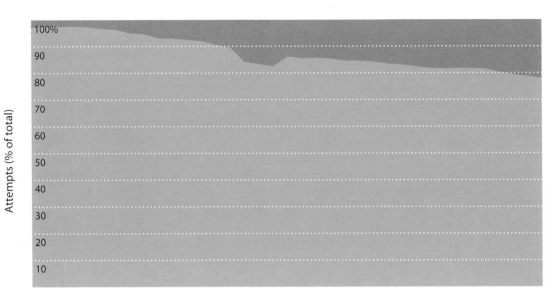

Attempts (% of total)

100%
90
80
70
60
50
40
30
20
10

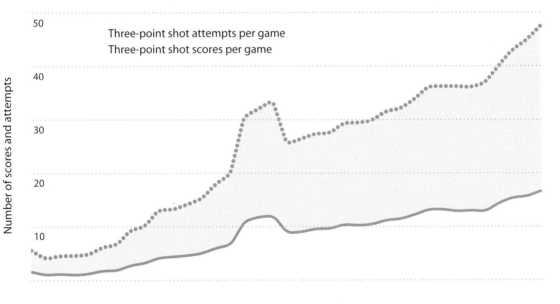

Number of scores and attempts

50

Three-point shot attempts per game
Three-point shot scores per game

40

30

20

10

1979-80 1981-82 1983-84 1985-86 1987-88 1989-90 1991-92 1993-94 1995-96 1997-98 1999-00 2001-02 2003-04 2005-06 2007-08 2009-10 2011-12 2013-14 2015-16

Losing: a winning strategy?

How basketball teams 'tank' to get better draft picks

In many team sports, having a terrible season has few benefits. You might well get relegated, and drop down a division. The knock-on effects can mean a team takes years to make it back to the top.

In America, instead of relegation, they have the draft. The lowest ranked teams get to pick earlier from the list of eligible college or amateur players, ensuring that over time, the league is evened up. It is essentially a socialist game changer.

In basketball, if a team gets a truly star player, things can turn around quickly. Basketball stars can carry a team like no other sport given that it's just five vs. five on the court.

This creates the incentive to throw games – to tank – in order to be lower down the rankings and therefore have first pick to get a higher ranked player.

How can you spot tanking? In the case of the Philadelphia 76ers, the general manager Sam Hinkie recently as much as admitted it, pointing out that teams don't improve gradually season to season, but completely revamp and then win big. Hence the strategy of building up early draft picks.

The 76ers' strategy is not winning them many matches – or friends. As LA Lakers owner Jeanie Buss said in an interview with ESPN: 'The teams that use tanking as a strategy are doing damage. If you're in tanking mode, that means you've got young players who you're teaching bad habits to. I think that's unforgivable.' Ironically, the Lakers were suspected of tanking the 2014–15 season, although by mid-season 2015–16 the team had an even worse win-loss record.

But how can we spot tanking? It's hard, clearly. The losing habit is hard to shake, whether legitimate or otherwise. Most players don't want to tank, as it only increases the possibility of some new hotshot player taking their place in the team.

One method is to look at season-by-season results. If a team goes from winning to losing then back to winning again, it is a pattern that fits with tanking, even if it is not conclusive.

Taking the win-loss ratio for each team, any time a team drops by a large amount from the previous season and then makes it up again the next could be under suspicion. However, this way of looking for tanking is far from perfect: it

doesn't capture teams that tank for two seasons in a row, or those that tank from a low base.

Another way to look at it would be consider every team's win ratio in a given season, and compare it to the average winning ratio of the previous two seasons and the following two seasons. This helps to spot more prolonged tanking. Any time there is a dip below 50 per cent – i.e. a team winning half the average games of both their previous two seasons and the following two – may fit the tanking theory.

This has happened 12 times in the NBA. The suspect teams and seasons are:

San Francisco Warriors in 1964–65
Philadelphia 76ers in 1972–73
Houston Rockets in 1982–83
Dallas Mavericks in 1992–93
San Antonio Spurs in 1996–97
Boston Celtics in 1996–97
Denver Nuggets in 1997–98
Atlanta Hawks in 2004–05
New Orleans Hornets in 2004–05
Miami Heat in 2007–08
New Jersey Nets in 2009–10
Charlotte Bobcats in 2011–12

Some of these seasons are now legendary. In particular, 1996–97 was something of a tank-off.

That year, the San Antonio Spurs were disrupted by injuries to three key players; but they were hardly rushed back into action, sitting out most of the season, and the Spurs ended up winning just 17 out of 82 games, having won 59 the year before. They won 50 the following season, having drafted star player Tim Duncan. Duncan then helped them to win five NBA championships.

The Spurs weren't the only team interested in drafting Duncan; the Boston Celtics also wanted him, but despite winning just 15 games, they didn't get the draft pick they needed. In a later interview, the coach of the Celtics at the time, Michael Carr, described the experience of sabotaging the season as going 'completely against your basketball DNA'.

Down and up again

12 basketball seasons that fit the tanking pattern

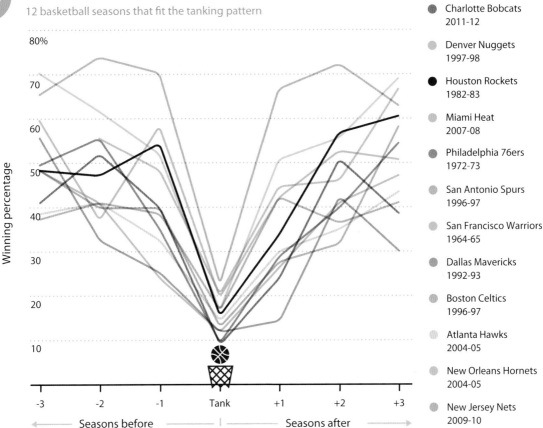

Winning percentage

80%

70

60

50

40

30

20

10

-3   -2   -1   Tank   +1   +2   +3

◀— Seasons before ——|—— Seasons after —▶

Charlotte Bobcats
2011-12

Denver Nuggets
1997-98

Houston Rockets
1982-83

Miami Heat
2007-08

Philadelphia 76ers
1972-73

San Antonio Spurs
1996-97

San Francisco Warriors
1964-65

Dallas Mavericks
1992-93

Boston Celtics
1996-97

Atlanta Hawks
2004-05

New Orleans Hornets
2004-05

New Jersey Nets
2009-10

One team accused of tanking is the Cleveland Cavaliers, who drafted LeBron James in 2003. That season they won just 22 games, but the team's year-on-year performance doesn't fit with the tanking pattern. In the three seasons before, the Cavaliers won 32, 30 and 29 games – that looks more like a poor team than a tank. This is the line where tanking and rebuilding meet. One can look much like the other.

Although the numbers suggest that the Houston Rockets tanked in 1982–83, the season that the Rockets are more infamous for tanking was the following year, when they won 29 games (they won 14 in 82–83). The team faded towards the end of the season and picked up their target draft pick: Hakeem Olajuwon, one of the greatest players ever, who led them to two championship titles.

In fact, the draft lottery, which randomly distributes the top draft picks amongst the teams that didn't qualify for the play-offs, was introduced following Houston's suspected tanking in 1984. The lottery has been modified several

times, but in essence it means no team can be assured of the picks they will get in the draft, which in theory should stop tanking. Yet it hasn't stopped some teams trying, clearly.

For every successful tank, there are lots of other teams that have tried and failed. And even star draft picks don't always guarantee success: James led the Cavaliers to the NBA title in 2016 after a couple of final disappointments – but only after a stint in Miami, where he won two titles first.

The perils of tanking:
- Current players feel undermined and team morale plummets
- The draft has too few stars
- The draft lottery means you may miss out on the pick you want
- The fans get upset and stop coming to games
- The team may be fined (in theory. In practice: no chance)

Bodysuit technology and swimming world records

Should we turn the clock back?

'Once we reach our limits, we go beyond them.' (Speedo, 2008, for the launch of the LZR Racer swimsuit.)

For once, the marketing was right. It may have been a misquote from Einstein (which in turn is possibly misattributed), but the full-body polyurethane swimsuits of 2008–09 changed what could be achieved in swimming forever.

In 2008 and 2009, as swimmers donned the new suits, records tumbled like never before: an incredible 110 swimming world records (50m pool) were broken by men and women. Some loved the suits, others were aghast at the way old times were made to look pedestrian.

Fina, swimming's governing body, decided it had had enough. It changed its regulations on fabrics and body coverage, which came into effect in 2010.

The chart shows the effect of the change in rules. In 2010 not a single swimming event world record was set, and just two in 2011. In the six years since the suits were outlawed, just 32 records have been broken.

In the blue-riband events of shorter freestyle, all the 50m, 100m and 200m records for men and women were set in 2009.

Not many men's records have been set since body suits were banned: just nine, including six in the three breaststroke events, one in the 1,500m freestyle, and one in the 200m medley.

More of the women's records have been broken, although it is noticeable that of the 23 records set since 2009; nine have been in long distance freestyle, and six in breaststroke.

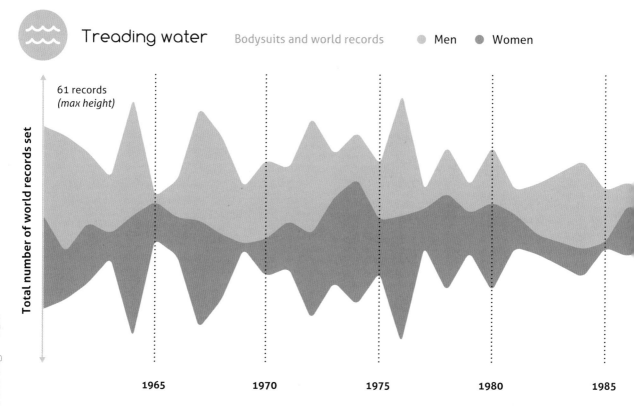

Treading water — Bodysuits and world records — ● Men ● Women

61 records (max height)

Total number of world records set

1965   1970   1975   1980   1985

Clearly, the design of the suits was less beneficial for longer distance swimmers and breaststroke. The suits were more helpful for the shorter freestyle events. This was partly because the extra buoyancy the suits provided gave a bigger advantage to heavier, more powerful swimmers. They were also more helpful in general for men: 15 of the 20 long course records survive from 2008 and 2009. For the women, it's seven out of 20.

For swimming, 2009 is the moment when the clocks stop, much like 1988 is for women's athletics.

The question is, what can be done? Unlike with athletics, where the problem is unproven drug use, Fina could easily erase any record set wearing one of the banned suits.

The dilemma is one of time. To erase the records now might seem too hasty. As Frank Busch, the US Olympic coach and director of swimming put it to *USA Today*: 'I think the suits were in some ways a great thing because they raised the bar. And these kids only know one thing: you raise it, we chase it.'

If the records are still standing in another 10 to 20 years, as the athletics ones are, then the case for erasing them becomes more compelling. Ironically, it then also becomes harder: some swimmers would have held and lost the revised record without even knowing it. Creating an accepted parallel history for two decades of sport is unprecedented. Even the Tour de France doesn't declare alternative winners for 1999 to 2005. It just has Lance Armstrong's name struck out.

How far out of reach are the records? This is where the bodysuit furore starts to look a little misplaced. Many of the records from 2008–09 are being chased down, with some recent times less than 10 hundredths of a second away. No record seems as untouchable as it did in 2010.

You can't erase the bodysuit records if they are going to be broken eventually; it's just a question of being patient. Swimming is just a little stuck – for now. It has reached its limits and then made it that bit harder to go beyond them.

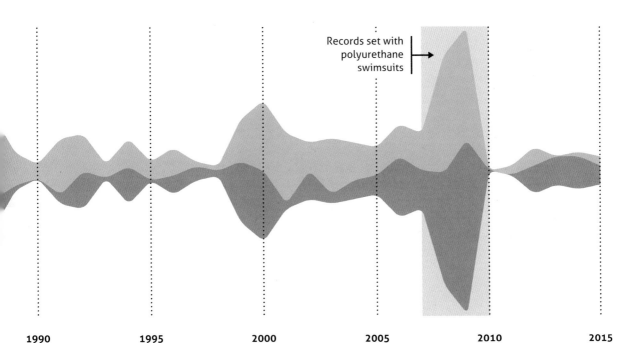

Records set with polyurethane swimsuits

1990 1995 2000 2005 2010 2015

Front crawl is a bit of a misnomer

The relative speeds of different swimming strokes make for interesting comparisons

No prizes for guessing that the 50m freestyle is the speediest race in swimming. Which is the slowest?

The 1,500m? The 400m medley (which involves an exhausting two lengths of each stroke)? In fact, if we take each of the main (long course) swimming world records and express them in terms of metres-per-second, it's the 200m breaststroke, by quite some distance.

Looking at the chart, breaststroke seems slow, but that's partly because of the now-banned bodysuits of 2008–09 (see previous chapter). The suits gave greater buoyancy and streamlining, which improved the backstroke and front crawl significantly, as well as assisting the butterfly stroke. The mechanics of breaststroke meant the suits were less help.

The chart also throws up a few interesting comparisons. Both men and women are slightly faster at 200m freestyle than 100m backstroke, which in turn is just slightly quicker than 50m breaststroke. The pace of the men's 800m freestyle is the same as the women's 200m freestyle. The men's 200m butterfly looks a little on the slow side, given that it is over 0.2 metres per second slower than the 100m, which looks like proportionately a big gap compared to other strokes. And for both men and women, you have to find events of 200m or above to find any record where the speed is slower than that of the 50m breaststroke.

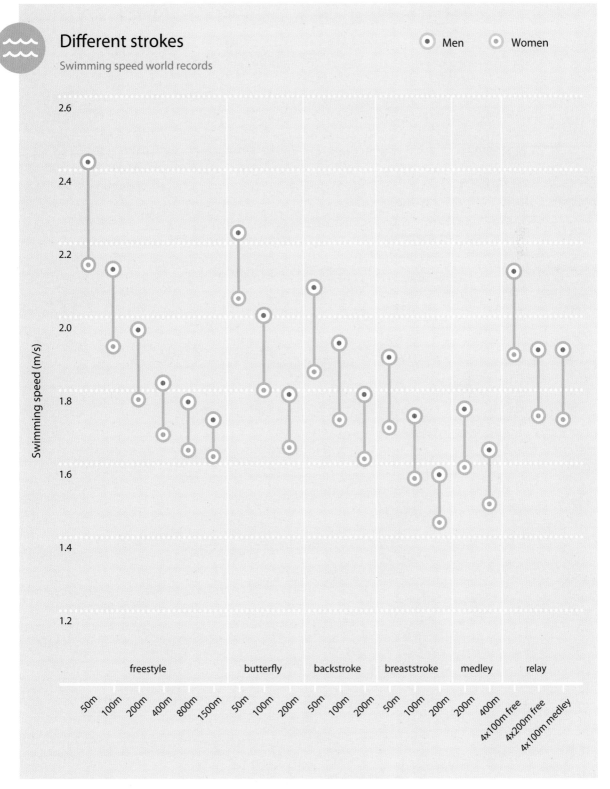

Different strokes

Swimming speed world records

● Men   ● Women

Swimming speed (m/s)

freestyle   butterfly   backstroke   breaststroke   medley   relay

50m 100m 200m 400m 800m 1500m 50m 100m 200m 50m 100m 200m 50m 100m 200m 200m 400m 4x100m free 4x200m free 4x100m medley

Winning isn't everything

Failure is necessary in sport, but it comes in many forms

Much of this book is about sporting success. That's because it can be readily measured and analysed, but failure is the other side of sport. There must be failure, or there can be no winners. But how do you measure failure? Here is a sport-by-sport guide to failure, including a Sports Geek alternative, when pundits point out the obvious.

Type	Sport / event	Stand-out failure	Reason
	American Football	Buffalo Bills, 1991-94	The Bills lost four consecutive Super Bowl games, the first by just one point after a last minute field goal to win the game just missed.
	Athletics	Trevor Misipeka, 2001	Trevor the Tortoise was supposed to be a shot putter. But at the athletics World Championships in Canada he was switched to the 100m as wild card entries were allowed in track but not field events. Cue a 14.28s 100m and cult status.
	Baseball	Chicago Cubs	When you get the nickname "lovable losers" you know something is up. The Cubs last won a pennant (the division, the step before the World Series) in 1945. They last won the World Series in 1908. Both droughts are a record, by some distance.
	Basketball	Karl Malone	Second in the all-time scoring list in the NBA, Malone has won many accolades but never the title that matters - the NBA Championship. His team, the Utah Jazz, played in the finals in 1997 and 1998, but came up against the Chicago Bulls, including Michael Jordan.
	Boxing	Reggie Strickland	Although losing hurts in any sport, imagine losing 276 fights. US fighter Strickland was a super-middleweight who competed in 363 fights from 1987 to 2005, recording just 66 wins (and 17 draws) along the way. No professional fighter has lost more.
	Cricket	Lance Klusener, 1999	The World Cup semi final between South Africa and Australia came down to the last over. Klusener hit eight of the first two balls and needed just one run to win. But a botched single resulted in a run out, and Australia went on to win.
	Football - Player	Roberto Baggio, 1994	One of best strikers in the world, Baggio had lit up the World Cup scoring five goals. But after a dull 0-0 final against Brazil went to a shoot out, the Italian missed his penalty, handing the South Americans the trophy.
	Football - Team	Brazil, 2014	Beaten 7-1 by Germany in the semi-final of the World Cup. The hosts were beaten at home for the first time since 1975, conceding five goals in 19 minutes in the first half.
	Golf - Round	Greg Norman, 1996	The most dramatic collapse ever? Norman led the Masters by 6 strokes on the final day, but lost to Nick Faldo by 5 shots at the end. Even Faldo felt sorry for him.
	Olympics - Summer	Eric Moussambani, 2000	Eric the Eel, from Equatorial Guinea, had never been in a 50m pool before the Sydney games. His time in the 100m freestyle was 1 min, 52 secs. The gold medal was won in 48.3 seconds.
	Olympics - Winter	Michael Edwards, 1988	Britain's Eddie the Eagle was a terrible ski jumper. At the Calgary games, he jumped 30m, compared to the winner who jumped over 120m. His look of permanent confusion and amateur demeanour set a new standard in lovable sporting losers.
	Rugby Union	Gavin Hastings, 1991	The World Cup semi final between England and Scotland was 6-6, 15 minutes to go, and Scotland were awarded a penalty right in front of the posts. Hastings was one of the most accurate kickers in the game, but he sliced the ball wide, and England went on to win.
	Tennis-Match	Natasha Zvereva, 1988	Beaten by Steffi Graf 6-0, 6-0 in the French Open final, in just 32 minutes. This remains the most one-sided final ever. In terms of the score alone, it can't be beaten.

Reason for Failure

⬤ Predictable calamity   ⬤ Colossal   ⬤ Choke   ⬤ Career / cumulative

Sports Geek alternative	Reason
Cleveland Browns	The Browns are the only NFL team to never host or play in the Superbowl. They have reached the playoff match prior to the Superbowl three times and lost them all.
Asafa Powell	Powell has twice set the 100m world record, and has run under 10s for the 100m more times than anyone else (84 times, 33 more than the next best). However, he has no individual gold or even silver medals in any of the major events.
Texas Rangers	The Rangers have been to two World Series and lost them both. They also made Alex Rodriguez the best-paid player in history, with a 10-year $252m contract in 2000, to little avail. The Rangers also filed for bankruptcy in 2010.
Elgin Baylor	Lots of great basketball players have fallen short of winning the NBA title, but Baylor stands out for going to eight finals and losing them all. The cruelest twist: in 1972 he retired due to knee injury, but his team, the Lakers, went on to win the title.
Johnny Greaves	While Strickland has the higest number of losses, the UK's Greaves had the sense to quit in 2013 after 100 fights, 96 of which were losses. His 4 per cent winning record makes Strickland's 18 per cent look positively accomplished.
Steve Waugh	Waugh, one of Australia's best captains, scorer of 32 centuries and over 10,000 test runs, is also the joint record holder for two test batting records that are better avoided: most dismissals in the 90s (10), and most ducks by a batsman (22).
David Trezeguet, 2006	Although it wasn't the final kick of the World Cup, the Frenchman was the only player to miss a penalty in the final shootout between Italy and France. By contrast, in '94 even if Baggio had scored, Brazil would still have won had they scored the next penalty.
Netherlands	Brazil's loss really takes some beating. However, in a way, the Netherland's three final losses and no wins is a more painful failure, although a cumulative one.
Jean van de Velde, 1999	Van de Velde led the Open by three shots on the last hole. Yet he hit his ball off the grandstands, into water, and then into a bunker. The resulting triple-bogey meant a play-off, which Paul Lawrie won.
Paula Barila, 2000	Also from Equatorial Guinea, also very slow, Paula 'the Crawler' finished the 50m freestyle in 1 min, 3 secs to do one length of the pool, compared to the gold medal time of 24.46 secs.
Phillip Boit, 1998	It's not a recipe for success: a Kenyan who hadn't seen snow two years prior to entering the Olympics as a cross-country skier. But Boit, despite coming in 20 minutes behind the winner at the 10k event, kept at his career in cross country.
Andrew Mehrtens, 1995	The World Cup final, 9-9, three minutes left. New Zealand fly-half Mehrtens received the ball in a perfect position for a drop kick but missed. South Africa went on to win in extra time, and it would take New Zealand another 16 years to win the trophy.
Bernard Tomic, 2014	Beaten by Jarkko Nieminen in Miami 6-0, 6-1 in just 28 minutes, a record as the quickest match ever (not including retirements). Tomic won just 13 points. His prize money worked out at $327 per minute, or $705 per point.

Secretariat stands alone

You don't have to be fast to win the Triple Crown, but it helps

You can hear the astonishment in race announcer Chic Anderson's voice. 'Secretariat is all alone! He's out there almost a sixteenth of a mile away from the rest of the horses!'

The Belmont Stakes, 1973. With around half the race run, Secretariat and Sham are neck and neck. By the end, the gap is over 30 lengths. Secretariat is so far ahead that it's hard to imagine the other horses were even in the same race.

That Belmont result is really an analogy for everything that has happened since. The Triple Crown, an unofficial title awarded to any horse that wins the three big cups of American racing, the Belmont Stakes, Kentucky Derby and Preakness Stakes, all have the same record holder: Secretariat, 1973.

Every year since, no horse has come close to winning in the same time. The nearest was Tank's Prospect in 1985

Triple Crown winners

The winners of America's 3 biggest races, and how far behind Secretariat they were

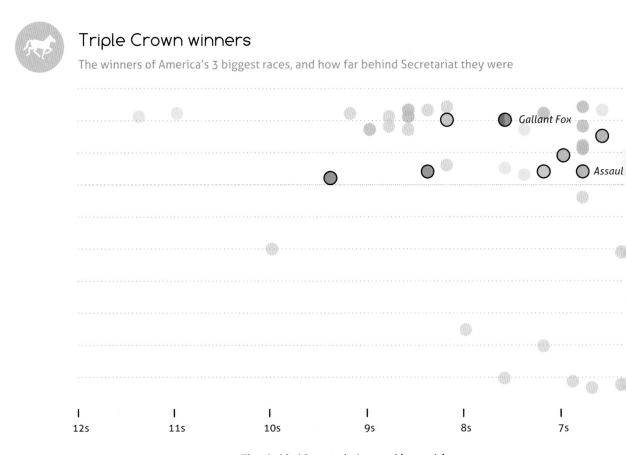

Time behind Secretariat's record (seconds)

and Louis Quatorze in 1996 at Preakness, both 0.4 seconds behind Secretariat's time. At the speed that horses run, that's still over two full lengths back.

It's as if Usain Bolt ran his times in the 1970s. And was around half a second faster than athletes now.

Instead, the Triple Crown has become so elusive that just the achievement itself is excitement enough. Forget the times, it's the titles.

In 2015 American Pharoah achieved the first Triple Crown since 1978. But his combined time was 11.8 seconds behind Secretariat (who happens to be his great-great-great-grandfather). In fact, in the last 90 years (since the three races have stuck to the same distance each year), American

Pharoah's 2015 combined time is the 53rd quickest.

Other Triple Crown winners have also been quite slow. The closest Triple Crown winner to Secretariat was Affirmed in 1978, less than three seconds behind Secretariat in each race, with a combined time 6 seconds behind.

With American Pharoah dubbed a super-horse, and destroying the field, it seems that we have recalibrated what time it takes to win these big races. Of course, the conditions change from year to year. But going on the winning times alone, American Pharoah was an average of 65m behind Secretariat. Those 1973 records look safe for a long time to come.

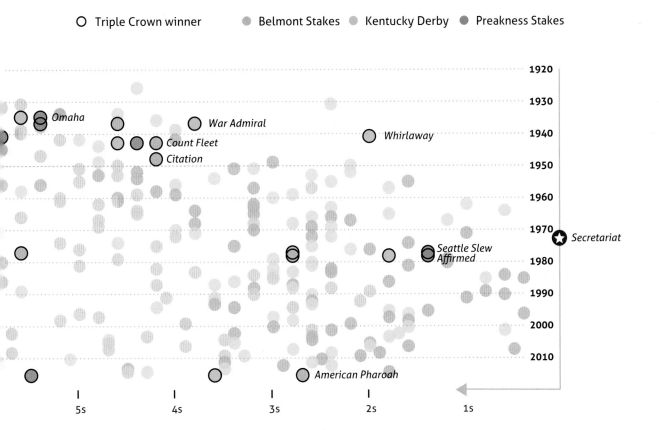

Are horses getting quicker?

A study suggests that the peak-horse theory is wrong

If you have read the chapter on Secretariat and the Triple Crown, you would be forgiven for thinking that horses are getting slower. No horse has come within a length of Secretariat's times since 1973. What further proof do you need?

Let's get one thing out of the way: Secretariat was an outlier – a nice way of saying a freak. There hasn't been a horse like him since. That's fine, but it doesn't tell us about horses in general.

Conventional thinking was that breeding had reached a 'selection limit' – in other words, there were no more improvements that could be made to the average thoroughbred through hereditary racing traits.

Patrick Sharman and Alastair Wilson of the University of Exeter published a paper in 2015 that blew much of that thinking away.

Taking into account all horses (not just winners) over a far greater number of races than other studies had used, and over a variety of distances, they allowed for ground conditions, field size and other factors.

Although their study was just of UK-based horses, the results were significant. While the evidence showed that horses hadn't improved much over longer distances, when it came to sprints, there was a marked improvement year on year, right up to the end period of the data (2012). As Sharman and Wilson put it: 'Ongoing improvement in sprint performance, not previously analysed, is much more rapid. Between 1997 and 2012, winning speeds for elite 6-furlong races have increased by an estimated 0.11 per cent per year, corresponding to an improvement in predicted winning time from 72.92 to 71.74 seconds. On good ground, a difference of 1.18s corresponds to over seven horse lengths, a distinct

Giddy up
The speed of elite race winners

● Sprint   ● Middle distance   ● Longer distance

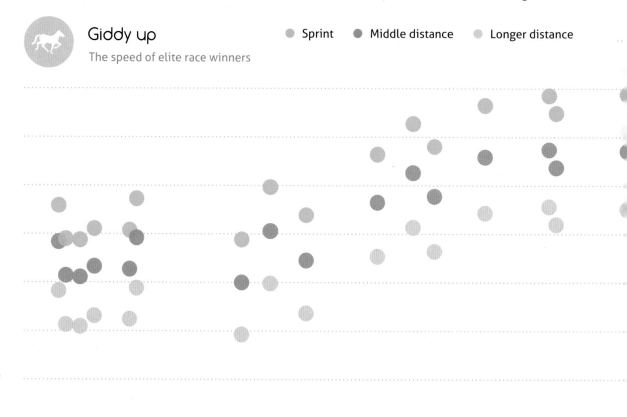

1850   1860   1870   1890   1900   1910   1920   1930

margin given that we calculated the average winning distance of 6 furlong elite races between 1997 and 2012 to be just 1.28 lengths.'

Quite a serious progression. Let's compare it to human sprinting. According to Sharman and Wilson, the speed of shorter (six furlong) elite race winners went from 18.1 to 18.4 yards per second over 1997 to 2012, an improvement of 1.65 per cent. Now let's take the top 15 fastest recorded 100m sprints of 1997 and 2012. For 1997, the average is 9.90; for 2012, it's 9.78, an improvement of 1.20 per cent.

It's a crude comparison, simply to point out that if we have noticed the change in men's 100m sprinting, why haven't we noticed the change in racehorses?

One reason is that we aren't so obsessed with times in racing. It's all about the betting and the winners. For the 100m, the goal is to race into the record books.

And in horse racing, there are far more variables. Athletics tracks differ, but not nearly as much as a racetrack. The distances are not uniform – the Sharman and Wilson dataset from 1850 has over 200 different lengths of race, including 42 they classify as a sprint (between 5 and 7 furlongs).

Why have horses improved in the shorter distances but not longer races? Sharman and Wilson suggest that for longer race formats, the possibility that we have hit the selection limit is quite real. Plus, breeders may have shifted their focus towards producing sprint horses. They also suggest that some speed improvements may come from non-horse variables like jockey position.

So we shouldn't give up on the possibility of seeing some records broken one day, even if Secretariat's times look out of sight for now.

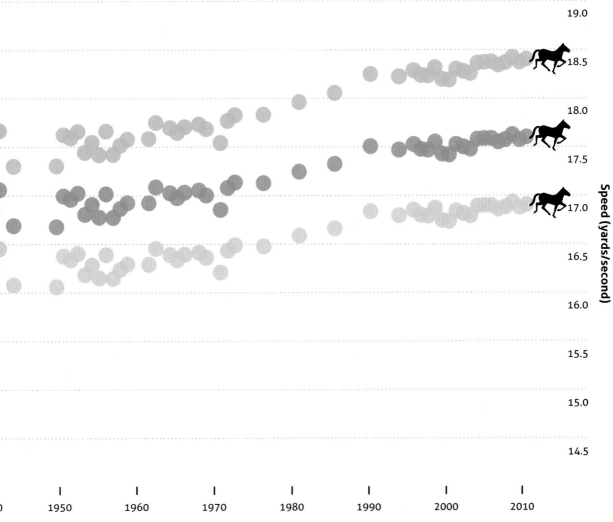

Canada has lost the NHL

The sport's spiritual home is in a wilderness

'The game commonly known as ice hockey is hereby recognized and declared to be the national winter sport of Canada.' (National Sports of Canada Act, 1994.)

Back in May 1994, when ice hockey was officially recognised as Canada's national sport, all was well. The Vancouver Canucks were about to take on the New York Rangers in the Stanley Cup final. In the 1993 final, the Montreal Canadiens had beaten the LA Kings four games to one.

Little did the Canadian lawmakers realise that Montreal's victory would be the last for a Canadian team. The Canucks lost a nail-biting deciding seventh game to the Rangers, 3–2. Since then, a Canadian team has contested the Stanley Cup just four times, only to finish runners-up each time. The wait is now over two decades.

What has gone wrong for Canadian ice hockey? How can a country which did so much to foster and create the sport in its current form have so little club success?

Part of the reason is quantity: there simply aren't enough teams in the National Hockey League based in Canada any more. In 1993, just before the national sports act was passed, Canada had 8 of the 24 NHL teams – a third of the league, with the other two-thirds of course in the US. That declined to 6 Canadian teams in a league of 30 for much of the early 2000s, although the relocation of the Atlanta Thrashers to Winnipeg has increased it to 7 in 30.

Canadian representation in the league has declined. Or, to be more precise, has stayed static while new franchises have set up shop in the US.

What prompted the move south? It was partly due to economics. The NHL pays in dollars – US dollars. And from the early 1990s to the early 2000s, the Canadian dollar fell in value from around US$0.85 to US$0.61. That meant that for Canadian teams, player salaries were harder to match every year compared to what the US teams could pay. By the time the currency trend reversed in 2004, it was too late – an

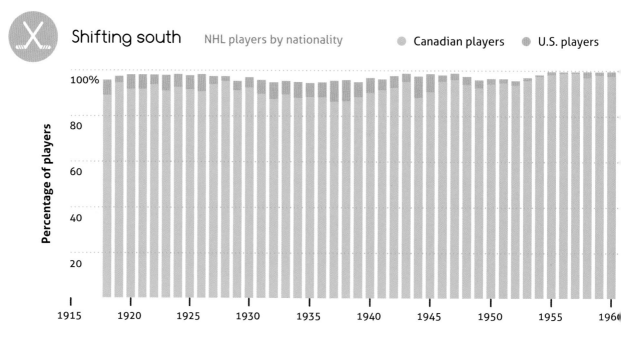

Shifting south NHL players by nationality   ● Canadian players ● U.S. players

Percentage of players

100%

80

60

40

20

1915 1920 1925 1930 1935 1940 1945 1950 1955 196(

extra eight US franchises had been created, while Canada had lost two.

Not only have the teams shifted south, so has player nationality. Until the 1970s, Canadian players used to make up over 90 per cent of the league. That has shrunk over the years to around 50 per cent, while US players now make up nearly a quarter. The other 25 per cent is from European countries: Sweden, Czech Republic, Russia and Finland.

Even if US teams were playing the big games, Canadians still had some interest due to the homegrown players involved. But that interest is now waning. The result is that soon the public will switch off. The 2015 Stanley Cup TV audience in Canada was 12 per cent lower than in 2014, with 2.39m viewers. Compare that to the deciding game of the Stanley Cup in 2011 – contested by the Boston Bruins and Vancouver Canucks – that was watched by an average of 8.76m Canadians. (That trend may be a worry for Rogers Communications, which paid $4.1bn for the NHL TV rights in Canada for 12 years in 2014.)

What can change this around? Success is the obvious answer. One or preferably two teams from Canada need to compete regularly in the playoffs. That should keep the public switched on. The question is how. The salary cap means teams can't buy their way to success like they can in other leagues. Creating new franchises in Canada would help – the league should be 32 teams, not the current 30, which creates an imbalance between the two conference divisions (with 16 teams in the Eastern Conference and 14 teams in the Western Conference).

Currently the NHL is still popular enough that Canadian fans come to games regardless of how their team is performing; but relying on the loyalty of fans may have made some Canadian team owners and managers complacent. A bit of fan dissent might help.

The other hope is that this is all just cyclical. Team fortunes can change. For Canada, the hope must be that it happens before the public switch off, and more teams move south.

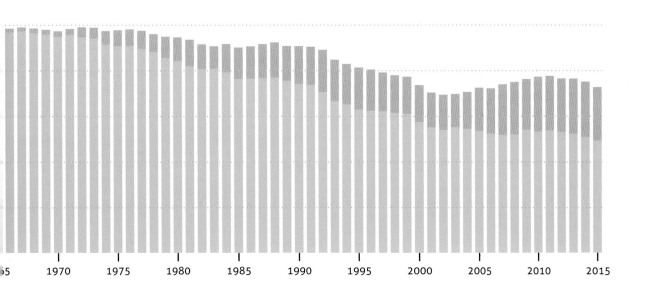

Fighting on ice

Ice hockey fights are on the wane, but not because players are nicer

Not many sports have a set of rules for fighting. Boxing does, of course. But in most other sports fighting is a sending-off offence.

Ice hockey is different. Many fans love them. Teams have 'enforcers', an unofficial but clear role. Fights are so ingrained in the game that the rulebook states:

46.1: Fighting – A fight shall be deemed to have occurred when at least one player punches or attempts to punch an opponent repeatedly or when two players wrestle in such a manner as to make it difficult for the Linesmen to intervene and separate the combatants. The Referees are provided very wide latitude in the penalties with which they may impose under this rule. This is done intentionally to enable them to differentiate between the obvious degrees of responsibility of the participants either for starting the fighting or persisting in continuing the fighting. The discretion provided should be exercised realistically.

And so on for another 2,000 words, encompassing 22 subsections, describing who is an aggressor, what can and can't be done with helmets and jerseys (both should stay on, at least to start with), where and how the fights take place.

But fighting is falling. The number of fights has dropped from around 1 per game in the 1980s and most of the 1990s, to close to 0.3 per game now.

Why? What's behind the clear drop off in fights?

Simply put: winning teams aren't fighters. The overall correlation of fights for any team to its win-loss record in the regular season is negative, at -0.27. That's not a hugely significant score: it would need to be closer to −1 to indicate a strong link between not fighting and winning. We can't say that teams win because they don't fight, nor can we say that they don't fight because they win.

We can say that fighting doesn't make a positive impact. The only team in the last 27 years (from the available data) to record a regular season points percentage of over 80 was the Chicago Blackhawks in 2013, a team that had only 16 fights, close to the bottom of the fights table. The Blackhawks went on to win the Stanley Cup that season.

Equally, the San Jose Sharks had a points percentage of just 14 in 1993, the joint-lowest in the league – and were top of the fights table with 81.

However, these are extreme examples – outliers – which could lead to the idea that fighting is detrimental to a team's chances. And while the negative correlation supports that idea, it's not conclusive by any means. Teams that fight more might have otherwise had an even poorer set of results, or perhaps fights happen when games are already lost. The numbers can't tell us those sorts of things.

However, it is clearly detrimental to players' health, often with tragic consequences. In 2011, three enforcers – Derek Boogaard, Rick Rypien and Wade Belak – died, two reportedly of suicide. Enforcers frequently suffer from depression, anxiety and substance abuse. Although there is an etiquette to hockey fights, there are personal effects far beyond the ice rink.

As the health impact becomes more widely reported, it may contribute to the fall in fighting. But in the end, the key factor is that, when it comes to results, fighting doesn't make much difference.

Hockey fights: on ice
Average number of fights per game

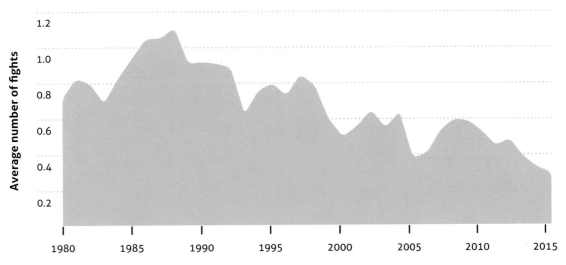

Fighting, not winning
Number of fights vs. percentage of points per season (1988-2015)

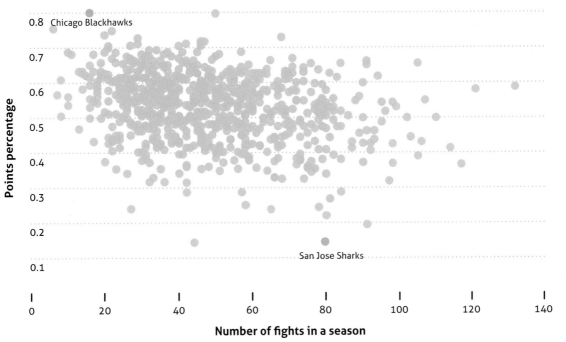

We're going to need a bigger net

Are bigger goaltenders ruining ice hockey?

The NHL has a problem. It's getting really hard to score goals, and it's not because the standards of shooting have fallen. It's because the goaltenders are just too big.

Let's look at the numbers. The goalies are now over 6 per cent taller than they were 30 years ago, and 14 per cent heavier. Plus the equipment goalies wear is bulkier.

The net? It's still 72 inches wide by 48 inches tall. The average wingspan of a goalie is now wider than the net.

As a result, the save percentage (shots saved out of shots faced) has gone up from 87.5 per cent in 1985 to over 91 per cent today. The average goals per game has fallen from around 4 in the 1980s, to 3 in 2005, to 2.6 now.

So why hasn't the NHL made the nets bigger? Well, goals aren't the only source of excitement. Not all fans like a high-scoring game. The game might be better improved by removing a player on both sides to create more space for players.

Making the net bigger would nevertheless be a simple enough solution. Some coaches have voiced the idea, such as Toronto Maple Leafs coach Mike Babcock, who said in 2015: 'The net's too small for the size of the goalies. Period.'

Not all coaches and players are in favour, however. Several have pointed out that the game is faster, the players shoot the puck harder than they used to (hence the need for extra padding) – and that other sports such as football don't suffer from having 1–0 scorelines.

So will the nets be made bigger? Unless there is close to a consensus, the net will stay as it is. But if the scoring rate continues to drop, the pressure will mount. The lowest goals-per-game average of recent years is 2.57 in 2003–04. If the rate fell below that, the calls for a bigger net would become very loud indeed.

In the way

Bigger goaltenders in the NHL mean more saves

Save %   Weight (lbs)   Height (in)

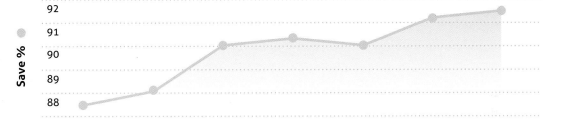

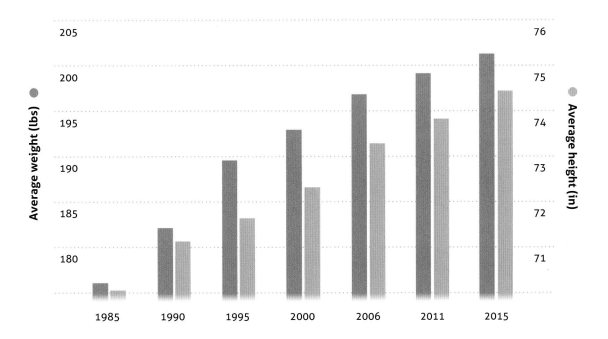

Big screen, small success

Aside from boxing, sports rarely translate well to film

What is it about boxing and the movies? Outside of Hollywood, the sport is in a fairly dire state. A lack of stars, too many dodged fights, split titles – as well as the periodic calls to ban it.

That's all forgotten when it comes to making movies. There are just 31 sports films that have made over $100m at the box office worldwide, and 12 of them are about boxing, including five of the top ten (see footnote for my definition of a sports film).

So what is it about boxing that works at the cinema?

There's history, for starters. Boxing has a strong film heritage, so actors and directors gravitate to the sport; but it boils down to two elements: the individual, and the underdog.

Team sports are that bit more complex in terms of storyline – there are all those other players and personalities that can't be ignored. Not so with individual sports. And if you want a tough, gritty underdog story, tennis and golf hardly fit the bill (though it has been tried).

That leaves boxing. It also gives serious actors a chance to show off some pain and suffering: see Russell Crowe, Will Smith, Clint Eastwood, Christian Bale and Robert De Niro.

Among sports films, boxing gets a high degree of critical and commercial acclaim. In the top 31 grossing sports films, the four best regarded are all boxing movies. *Rocky I, Million Dollar Baby* and *Creed* all have a rating on the online film database IMDb of 8 and above, and *The Fighter* is on 7.9. (Probably the best boxing movie of the lot, Martin Scorsese's *Raging Bull*, rated 8.3 on IMDb, doesn't make the list, with just $23m grossed worldwide.)

After boxing, American football also does well. *The Blind Side* is in second place on the top-grossing sports list with over $300m and a good IMDb score of 7.7. Overall, American football films make up seven of the top 31.

However, the top box office sports film of all time with over $350m is not about boxing or gridiron, it's *The Karate Kid* (the 2010 remake, rather than the 1984 original). Perhaps it is no coincidence that violent sports do well at the cinema; violence in general seems to work on film.

But despite the popularity of live sport, sports movies aren't a big deal in the overall scheme of Hollywood. *The Karate Kid* may sound like a big success, but that puts it in the all-time list at 272nd place, just below *Crocodile Dundee* and just above *Home Alone 2*, according to WorldwideBoxoffice.

The next best-represented sport is auto racing, with three films in the top 30, although the films are poorly rated. *Herbie: Fully Loaded* gets a measly 4.7 on IMDb. The best-regarded motor racing films of recent years made little impact at the box office. *Senna*, the documentary about the Brazilian legend, gets an IMDb rating of 8.6, but made just $12m; *Rush*, the polished James Hunt/Nikki Lauder biopic, is rated 8.2 but made $98m, just outside the top 31.

Some sports seem woefully under-represented at the cinema. Basketball's only entry in the top 30 is *Space Jam*, a feeble cartoon-NBA basketball adventure. It gets only a 6.3 rating on IMDb. But it has Michael Jordan in it – and back in the 90s that was enough to guarantee success, given his popularity. It grossed a quarter of a billion dollars worldwide. Commercially, it's the only big breakthrough basketball film. With a bit more critical success, but just outside the top 30 with $90m, is *White Men Can't Jump*, rated 6.6 on IMDb. Again, it was made in the 1990s. There has been no major basketball film for the best part of two decades. Even the most critically acclaimed basketball film was made in 1994 – the documentary *Hoop Dreams*, which has an 8.3 rating on IMDb but grossed only $12m worldwide.

Baseball's lack of big movie success is equally puzzling. The sport sits deeper in the American psyche than other sports and has attracted great writers such as Don DeLillo (*Underworld*) and Philip Roth (*The Great American Novel*). Perhaps Chad Harbach's bestselling novel *The Art of Fielding* from 2011 will provide the material for a truly breakthrough baseball film. But up to now, the sport hasn't set the movies alight. Only *Moneyball* and *A League of their Own* sneak over the $100m mark. Not a single baseball movie gets a rating above 8 on IMDb.

Football (the soccer variety) fares even worse. Despite being the world's biggest game, the sport has yet to create a major global film: 2003's *Bend it Like Beckham* (rated 6.7) is still the top grossing soccer film of all time, with just $75m. It's also the top grossing British sports film, too, unless you count *Rush*, which is a British-German co-production.

Other British sports films have not fared well at the box office, whether they have pandered to a US audience (such as *Wimbledon*, which made $41m) or had a largely domestic

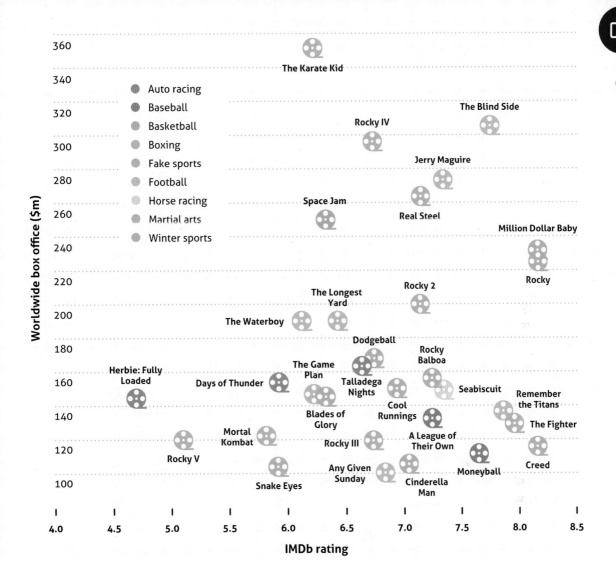

Worldwide box office ($m)

●	Auto racing
●	Baseball
●	Basketball
●	Boxing
●	Fake sports
●	Football
●	Horse racing
●	Martial arts
●	Winter sports

The Karate Kid

The Blind Side

Rocky IV

Jerry Maguire

Space Jam

Real Steel

Million Dollar Baby

Rocky

Rocky 2

The Longest Yard

The Waterboy

Dodgeball

Rocky Balboa

Herbie: Fully Loaded

The Game Plan

Days of Thunder

Talladega Nights

Seabiscuit

Remember the Titans

Cool Runnings

Blades of Glory

Mortal Kombat

Rocky III

A League of Their Own

The Fighter

Rocky V

Any Given Sunday

Creed

Snake Eyes

Cinderella Man

Moneyball

IMDb rating

appeal (*The Damned United*, which made $4.2m, less than production costs). Both films are still comfortably outgunned by *Chariots of Fire*, which made $57m back in 1981.

Filmmakers have tried their hand at smaller, less obvious sports too. *Cool Runnings*, a film about the Jamaican bobsleigh team, and *Blades of Glory*, an ice skating comedy, are some of the more surprising entries in the top 31 list. And for almost every sport, there's a notable attempt: golf has *Tin Cup*, tennis has *Wimbledon*, rugby has *Invictus*. In general, they have made little impact. For sport at the movies, nothing works quite like boxing.

Sports Geek's definition of a sports movie:

For inclusion in my list, sport has to have a key role, whether as the backdrop to the story or as a central plot device. Sport has to be a defining feature, such as in *Jerry Maguire*, even if

the main two characters don't play.

Not included in the list are films that feature a sport briefly; *American Pie* (which would get into the list on box office takings) has a good Lacrosse scene, but is by no means a sports film.

Kids films that have a sporting connection such as *Cars* and *Turbo* have been excluded as they are aimed at a different audience.

Most martial arts films have been excluded: they aren't really about the sport as such. *The Karate Kid* is included as it is about the sport – there is a tournament, after all – rather than just using a martial art as a way of beating up baddies. For that reason, *Karate Kid II* would not make the list as the fighting is not about sport.

In this list, *Dodgeball* counts, despite being a comedy and featuring (what was at that time) a spoof event.

Going down, sooner

Heavyweight boxers hit the deck far earlier than other fighters. Why?

The undeniable excitement of a boxing match is knowing that it could end at any moment. One punch is all it takes.

That also makes it a very fixable contest. Unlike a team sport with a set amount of time, or one with an incremental scoring system (such as tennis), a boxer just needs to drop his guard for a second. The simplicity of throwing a fight also makes spotting a fix incredibly hard.

Which brings us to the heavyweights.

Sports Geek has analysed every title fight across all divisions of the four main boxing belts – WBA, WBC, IBF and WBO – since the formation of the WBA in 1962.

Each fight can end in a number of different ways: knock out, technical knock out, retirement, disqualification, points decisions and so on. So let's look at the fights that ended in knock out, technical knock out (where a boxer is deemed unable to carry on by the referee) and retirement (when a boxer can't start the next round). For simplicity, let's call them stoppages.

The overall rate of stoppages for all title fights has fluctuated between just under 40 to just over 60 per cent of fights over the decades. Since the late 1980s it has trended downwards. That might be because boxers aren't as good finishers as they were before, or because they are fitter. But what is more interesting is the differences between weight divisions.

As the chart shows, the stoppage rate tends to go up as boxers get heavier. Super flyweight fights are stopped in 43 per cent of fights, the lowest of any division. The highest is heavyweights with 59 per cent.

Why might this be? The obvious conclusion is that punching power goes up with weight more than the ability to take a punch. Or, it might be that heavier boxers are less able to dodge a punch. Both would contribute to the pattern of stoppage rates.

So far, so expected.

So now let's look at the average round in which the stoppage takes place. For this analysis, it doesn't matter if it was a 12 or 15 round contest – this is the average round

when a stoppage fight (not points-decision) ends.

There is a remarkable similarity for the stoppage round for every division except for one: heavyweight. On average, a heavyweight stoppage is in round 6. Every other weight division has an average stoppage round of 9 or 10.

This is a big difference. What possible reason could there be for the heavyweights to be knocked out so much earlier than every other weight division?

Well, it could be that they are hit so much harder. But then we would expect the average stoppage round to come earlier and earlier as you move up through the weight divisions. Instead, light heavyweight and cruiserweight, the two divisions below heavyweight, on average are knocked out in round 9 or 10 (see chart).

Perhaps there are more mismatches than other divisions? The greater weight variance does mean there are more David vs. Goliath fights. However, they don't result in more stoppages.

The range of a typical boxing weight division is 8 lb. For heavyweights, a difference of 8 lb or less in weight ends in a stoppage in 60 per cent of title fights. A difference higher than 8 lb ends in a stoppage 59 per cent of the time. That's basically no difference.

Heavyweight fights that stop in rounds 1 to 6 have an average weight difference of 14 lb. Heavyweight fights that end in rounds 7 and above have an average weight difference of 13.3 lb. Whichever way you look at it, big weight differences don't seem to be to blame.

So what's left? It could be that fighters are going down early when they are told to.

As discussed earlier, it's not technically hard to fix a fight. There are other ways too: a points decision could be fixed – but it's harder as there are judges to get on board. It has happened: the Seoul Olympics is probably the clearest example, when Roy Jones Jr was awarded silver despite clearly out-punching his Korean opponent, Park Si-Hun.

But without proof of fixing, we can't be sure. Perhaps there are some other factors that would make heavyweights get stopped earlier. Maybe heavyweights are less fit than

It's a knockout

Stoppage rate and average stopping round by weight division

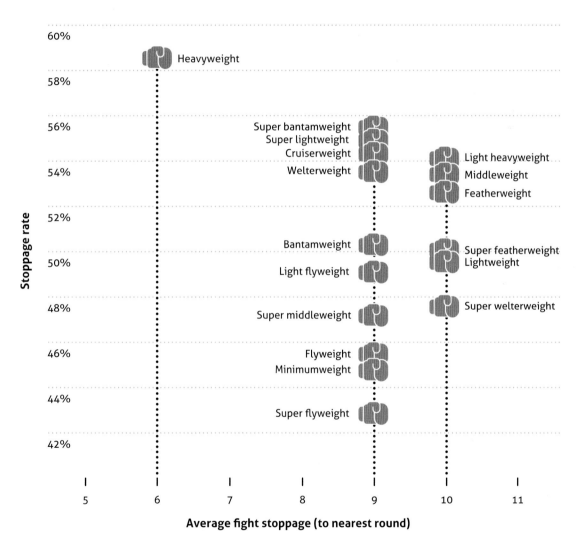

lighter boxers, or there are more mismatches in the division because the higher sums of money attract a wider range of fighters.

Whatever the reason, the average heavyweight stoppage round is clearly an anomaly.

The greatest boxer ever you (probably) never heard of

Pound for pound, it's Archie Moore

Stats and boxing don't exactly go together. How can the brilliance of Muhammad Ali or the brutality of Mike Tyson ever be expressed as a number?

Everyone has their favourites, but debating who is the greatest just in terms of basic won-lost records is little use for comparing across generations or weight divisions. Add style into the debate, and trying to find a consensus is futile.

But there is a statistical method for working out who is the greatest boxer of all time. The website Boxrec uses a formula to work out each boxer's rating. It's complicated, but essentially it takes every fight in its database and awards boxers points according to the type of result, the judges' scores (if applicable), and changing points as they move divisions, as well as the ranking of the two boxers before the fight.

What emerges is probably the best mathematical model of pound for pound boxer rankings. Pound for pound is fighting-speak for comparing boxers of different weights.

So who is top of the all-time list? The clear winner is Archie Moore, a light-heavyweight who fought from 1935 to 1963, with 2,534 points.

Not Muhammad Ali? Well, Ali is the top-ranked heavyweight of all time, and third overall on 1,991 points. Second was the great middleweight Sugar Ray Robinson, with 2,395. Current welterweight Floyd Mayweather is fifth with 1,755 points.

Who was Archie Moore? His boxing record of 185 wins (including 131 knock outs), 23 losses and 10 draws is formidable. Perhaps more impressive is the fact he was hospitalised twice, once in 1941 for an aggressive ulcer followed by appendicitis, and again in 1955 for a heart condition, and both times told he would never fight again.

He defied those odds, and went on to fight in the heavyweight division as well as become light heavyweight world champion.

Moore fought both Rocky Marciano and Muhammad Ali and lost to both. In the title fight with Marciano, Moore knocked the champion down onto the canvas for only Marciano's second (and last) time in his career, but lost in round 9.

The fight in 1962 with Ali – then still known as Cassius Clay – was Moore's penultimate fight, and footage shows a 40-year-old man who looks skilled and determined, but too small, too slow and too unfit to stand a chance. He was knocked out in round 4.

According to a *New Yorker* profile of Moore in 1961, 'because his build is undistinguished and countenance unscarred, there is nothing about his appearance that hints at the violent nature of his trade, and he affects a wispy bebop goatee that gives more the look of a jazz musician than of a fighter.'

Moore was clearly a great talker. The *New Yorker* reported Moore saying: 'I'd rather use six little words than one big one... There is, I fear, that one chink in my armour. I am inclined to waste words. In fact, I throw them away. It is my only excess.'

Moore also had a sense of perspective and humility that is often lacking in boxers. In 1960, towards the end of his career, he wrote in his autobiography *The Archie Moore Story*: 'Some people say it's great when a man retires undefeated champion. But the world champion means I must be beaten. A champion should fight to the finish and go out with his hands cocked just as he came in. It's the proper exit and I think it may be mine.'

Pound for pound: the greatest boxers of all time

Top ranked boxers according to the BoxRec scoring system

- flyweight
- bantamweight
- featherweight
- super featherweight
- lightweight
- super lightweight
- welterweight
- super welterweight
- middleweight
- light heavyweight
- heavyweight

BoxRec score

Archie Moore

Sugar Ray Robinson

Muhammad Ali

Joe Louis

Floyd Mayweather Jr.

Harry Greb

Ezzard Charles   Carlos Monzon

Sam Langford   Dick Tiger

2500 —
2250 —
2000 —
1750 —
1500 —
1250 —
1000 —

1880 1900 1920 1940 1960 1980 2000 2020

Years active

Paid to play, or paid to smile?

The best-paid sports stars get their money in very different ways

For the top sports stars, which is the bigger source of income: salary or sponsorship? Given the huge number of adverts featuring sports stars, you would be forgiven for thinking that sponsorship would win out. In fact, sponsorship is not even close. According to Forbes, the top 100 best paid sports people in the world in 2014–15 (the 12 months from June to June) earned over $2.3bn playing their sport. Their total endorsements were $836m – 35 per cent of their on-field earnings.

It's a complicated picture, though. The amount of endorsements – or sponsorship – that the top stars get has little direct relationship with total winnings, or salary, and more to do with the type of sport they play. Of the top 100, only 18 earn more from endorsements. Of those 18, 13 are in individual sports; it's easier for a golfer or tennis player to market themselves than it is to stand out from the rest of a team.

Individual stars, however, have a far less reliable source of income. If, say, Rafael Nadal gets injured and misses a few big money tournaments, that's his misfortune. He doesn't have a club still paying him $100,000-plus per week, like a star player in a top team does.

Of the five team players that get more endorsement than salary, three are in the US where heavily marketing players from team sports have a long history. They are basketball stars LeBron James, Kevin Durant and Kobe Bryant. The others are India cricket captain Mahendra Singh Dhoni, whose job is arguably the most stressful and scrutinised in world sport, and Brazilian footballer Neymar who plays for Barcelona.

Among the endorsement-led money earners, Tiger Woods and Roger Federer stand out with over $50m each in deals, far more than their prize money. However, in terms of multiples, that's nothing compared to Usain

Bolt, whose off-track earnings of $21m were 1,400 times his winnings, which are the lowest in the top 100 at just $15,000.

Football stars Cristiano Ronaldo and Lionel Messi both do well off the pitch with sponsorship of over $20m each, but their salaries are so extraordinarily high (around $52m) that their sizeable endorsements are only around half their wages.

Ndamukong Suh, a defensive linesman for the Miami Dolphins, who is a new entrant on the list with a salary of over $38m, is the highest-ranked sportsperson to earn less than a million in sponsorship deals.

There are only two women in the top 100, both tennis players. Maria Sharapova is the best-paid female in 26th overall place, with $23m in sponsorship. Serena Williams, at 47th overall, had the most winnings as the world number one, but earned $13m in endorsements.

Tennis is the surest route to sporting riches for women, with seven of the top 10 female sports earners according to Forbes. Nascar racer Danica Patrick is the top-paid non-tennis player, with $14m.

Overall, the sport that stands out the most as lacking endorsement is boxing. Floyd Mayweather is top of the total pay list, with over $300m, and his rival and opponent in the biggest pay-day fight of all time, Manny Pacquiao is second on the list with $160m. Both get far less ($15m and $12m) from endorsements. That's an improvement on the year before, when Mayweather got zero sponsorship.

His bad boy image and history of domestic violence hardly helps. And clearly, he doesn't need the money. But boxers in general aren't doing well in endorsements. It's slim pickings compared to other top sports stars. Boxing is certainly a popular sport – it was one of the first that was a viable pay-per-view proposition, after all, and plenty of boxers have been in adverts in the past. However, the sport's reputation has fallen, hampered by a confusing number of world titles and a dearth of compelling stars.

So in terms of pay, what has changed over the years? When we compare the data to 1997, which is the earliest salary/endorsement breakdown available from Forbes for the top sports stars (top 40 only), some patterns remain. Golfers and tennis players were able to attract endorsements well ahead of their winnings, whereas team players had the majority of their earnings from salary. One player makes it into both the 1997 and 2015 lists: Tiger Woods.

The stand-out figure from 1997 was Michael Jordan, but he was an exceptional sports star in every way. Even his salary from back then would be in the top 10 today; add on his endorsements and his total pay of $78.3m would have put him in overall fourth place, just behind Cristiano Ronaldo. At the time of going to press, Cristiano Ronaldo was the highest-paid sports star in 2016.

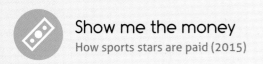

Show me the money
How sports stars are paid (2015)

● Salary/winnings ○ Endorsements

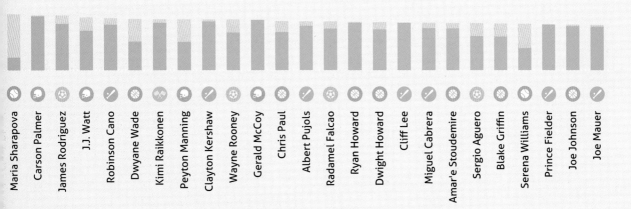

Maria Sharapova · Carson Palmer · James Rodriguez · J.J. Watt · Robinson Cano · Dwyane Wade · Kimi Raikkonen · Peyton Manning · Clayton Kershaw · Wayne Rooney · Gerald McCoy · Chris Paul · Albert Pujols · Radamel Falcao · Ryan Howard · Dwight Howard · Cliff Lee · Miguel Cabrera · Amar'e Stoudemire · Sergio Aguero · Blake Griffin · Serena Williams · Prince Fielder · Joe Johnson · Joe Mauer

It's all downhill from here

The gap between the fastest skiers' times has started to widen again. Why?

The men's and women's downhill skiing of the 2010 Vancouver Olympics were both gripping spectacles. Yet they could not have been more different.

In the men's event, the top three skiers were separated by just 0.09 of a second – the closest race in Olympic history. Didier Défago of Switzerland won the gold in 1 minute 54.31 seconds. The gap to Bode Miller of the US who took bronze was the equivalent of just 2.5m over a 3km course. Even the 10th placed skier was close: Michael Walchhofer of Austria finished just 0.57 seconds behind Défago, a tiny gap.

If the men were squeezed tight together, the women were stretched along the course – some literally. The winner, Lindsey Vonn of the US, was 0.56 seconds ahead of her compatriot Julia Mancuso who won silver – almost the same gap as the top 10 men. The gap to the third-placed woman, Elisabeth Görgl of Austria, was 1.46 seconds, a bigger difference than the top 21 men.

This isn't because women skiers are slow. Nor is it because they have inferior equipment. It's because the women's course was incredibly tough. As Associated Press reported afterwards, German veteran Maria Riesch (who came eighth) described it as the most difficult course she had skied – ever.

Due to bad weather that had tightened the schedule, the women were not able to practise a run on the whole course in one go, and there were six crashes out of 45 skiers – several that looked horrific. The course was changed for the combined event later in the Olympics.

While Vancouver 2010 is a lesson in how the different courses can affect performance, the overall trend in World Cup skiing (the top international circuit) has been for tighter and tighter races.

A look at the two charts shows this. The dots represent all the downhill World Cup races since 1968, and show how far behind the winner the 10th placed skier was, in percentage terms. (We have used percentage rather than absolute time difference, as otherwise it would not account for the length of races.)

Both the men's and women's events show a marked improvement, with the 10th skier getting closer to the first on a consistent basis up until around 2005. Then the gaps start to widen again, as shown by the trend lines.

Why might this be? There are several factors involved in a skiing race: the weather, the equipment, the standard of the skiers, and the toughness of the course. As Vancouver showed, a tough course stretches the field. Weather will also play a big part.

Over the first three-and-a-half decades of World Cup skiing, the narrowing gap between tenth and first can't be from weather. It's most likely to be from improved skiing standards, and better equipment.

So what is behind the recent increase in the gap? It could be that skiing standards have dropped, which seems extremely unlikely. It could be due to consistently worse weather – also unlikely. So either it's tougher courses, or equipment.

Although there is some variation in the courses, that's not the main factor. The significant change is the rules on equipment. For the 2003–04 season, the International Ski Federation (FIS) announced that downhill skis would have to be a certain minimum length, and also restricted the shape of the ski, which can make for sharper turns. The result has been to reverse the tightening gap between the top skiers. When the FIS introduced further rules on the curvature of skis for the 2012 season, Bode Miller described the change to ESPN as 'a complete joke'. Miller added: 'from 1999 to 2003 was the peak of equipment in ski racing. Since then, it's all gone in the wrong direction'. The numbers back that up.

The skiers might be unhappy, but is this a good thing for the audience? The excitement of the Vancouver men's race might suggest that closer races make for better viewing. But by that measure, so do dangerous courses, and no-one is advocating that after the experience of the women's event.

Catching up

How far behind is 10th place?

 Men  Women

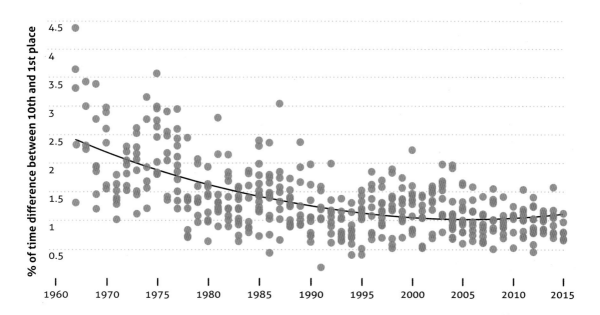

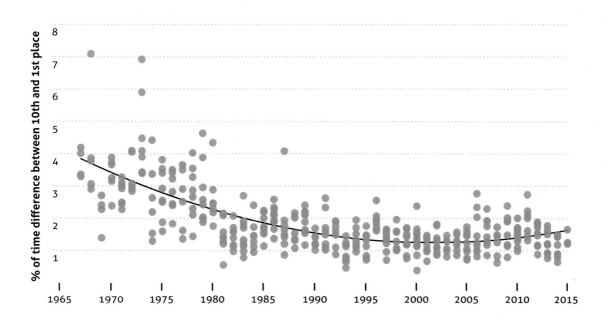

Where Andorra is bigger than China, for now

What type of skiing you do largely depends on the country you are from

Alpine, which covers slalom to downhill, is what most people think of when it comes to winter sports. The International Ski Federation (FIS), however, recognises several other categories, including cross country, ski jumping, freestyle skiing, telemark, snowboarding, speed skiing and even grass skiing. Within the professional ranks, which type you do is largely determined by which country you are from.

Norway dominates the world of competition cross country, providing nearly a third of all the competitors in that category. However, it has fewer alpine skiers than several other countries, including Canada and France. The countries with the greatest alpine representation are Japan, the US and Italy, each with over 10 per cent of the total category.

Countries tend to produce certain types of winter sports competitors. Japan has a lot of snowboarders and freestyle skiers. Germany and Slovenia seem to produce a high number of ski jumpers: in fact Slovenia has more ski jumpers than alpine skiers. Whereas in Spain, pretty much everyone is skiing alpine – over 80 per cent.

Some country comparisons defy expectations. Iran has more alpine skiers than Finland. Poland has more ski jumpers than Italy and the US combined. Norway has fewer snowboarders than Turkey. Andorra (pop. 85,000) has more alpine skiers than China (pop 1.3bn).

It's worth bearing in mind the size of the talent pool when the next Olympics comes around. Korea has a total of 257 registered competitors with the FIS, and China has 333. There are 13 countries with over 1,000 competitors. The next hosts are PyeongChang in 2018, and Beijing in 2022. If Korea and China want to celebrate some medal success, they need to get their skates on.

1,112

Canada

2,124

United States

1,3

Fran

25

Sp

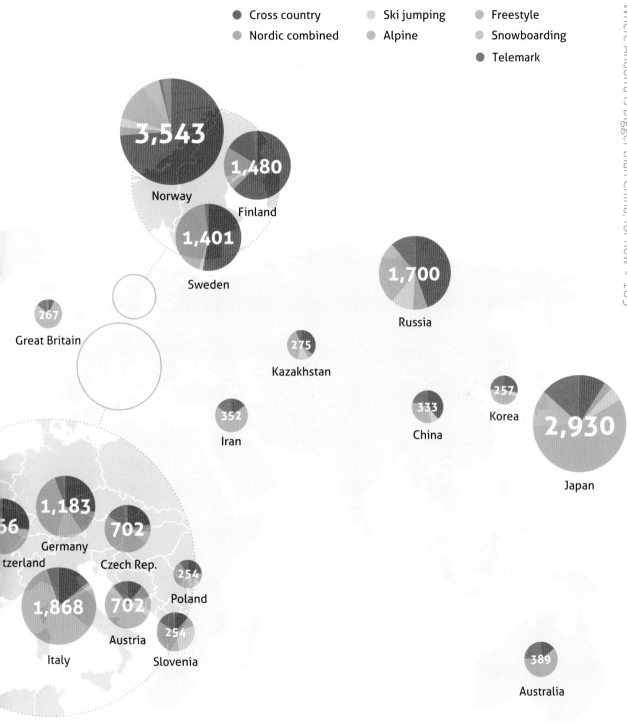

Cross country
Nordic combined
Ski jumping
Alpine
Freestyle
Snowboarding
Telemark

3,543 Norway

1,480 Finland

1,401 Sweden

267 Great Britain

1,700 Russia

275 Kazakhstan

333 China

257 Korea

2,930 Japan

352 Iran

1,183 Germany

6 tzerland

702 Czech Rep.

254 Poland

1,868 Italy

702 Austria

254 Slovenia

389 Australia

Up in the air

Accuracy isn't as important as it once was

The 2015 Grand Final of the ANZ Championship, the biggest netball league in the world. Less than three minutes remained on the clock, and the Queensland Firebirds were four points behind the New South Wales Swifts, 56–52. The Swifts were on the verge of an upset victory.

Then the Swifts made a couple of errors, and Romelda Aiken, the Firebirds Goal Shooter, netted four times without reply to draw the scores level: 56-all.

With only 15 seconds left in the match, Firebirds Goal Attack Gretel Tippett had the ball, a few yards back from the net. Her teammate Aiken was marked. She was out of time. The Fox network commentators called it correctly. 'Tippett doesn't want to take the shot,' said Kelli Underwood. 'She'll have to,' replied Liz Ellis, her voice straining to be heard above the screaming crowd. Tippett paused, then took the shot.

Up to that point, Tippett had scored just nine from 16 shots in the match – a lowly percentage of 56, below her season average of 64 per cent. Other Goal Attacks in the league typically score at around 80 per cent or higher. She was having a bad match: she had had only one shot in the fourth quarter, and had missed that. This was not your go-to player in the dying seconds.

The ball cannoned three times on the rim, and went in. The Firebirds celebrated like crazy, winners of the closest and most exciting final for years, having come up short in the previous two finals.

If one match sums up how netball has changed, the 2015 Grand Final of the ANZ league in Australia and New Zealand is it.

You might think that accuracy is crucial in netball. Only two players, Goal Attack and Goal Shooter, can take aim and score. Given that they shoot from a stationary position, it would seem obvious that the most accurate shooting team will win. Think again.

In 2015 the Firebirds not only won the final, but lost only one match all year. They also happened to have the lowest shooting accuracy of any of the 10 teams in the regular season – just 77 per cent, compared to 80-plus per cent accuracy of eight other teams.

Back in 2013, the Southern Steel won 6 of their 13 games. Yet their shooting was on target over 90 per cent of the time, the only 90-plus percentage ever recorded in the league. The season before, 2012, was the apex of accuracy: quite simply, winning teams tended to get more of their shots in. Meanwhile, total shots told you nothing about which teams won.

By 2015, the roles were reversed. As the charts show, accuracy was unrelated to success; shot attempts were a much better guide. So if accuracy is not so important, what makes netball teams win?

The answer isn't obvious. Statistics such as the number of interceptions and turnovers don't correlate to winning records that consistently when measured season by season. Some years there is a strong link between a stat and results: in 2010 for example, teams that lost the ball a lot (turnovers) won fewer games. In other years, there's little or no discernible relationship.

What has changed is the number of shots. The total number of shot attempts in a season has risen from around 8,300 to over 9,000, which works out at about an extra 10 per game. Netball has switched from accuracy to volume. It's less of a problem if the ball doesn't go in if another chance comes along in quick time.

In 2015, as well as being bottom of the league for accuracy, the Firebirds had 924 attempts on goal, which was the highest. That's 90 more than the league average.

Does that mean that teams with the highest shot count win? Not always. In seven years of the ANZ Championship, the team with the most goal attempts has had the best regular season record four times; but they have also been the worst. Meanwhile, the league leaders in accuracy have never had the top regular season record.

What this shows is a sport in flux: tactics evolve, whether it's shooting more often, or trying to steal the ball more quickly, and then other teams work out what to do. Like netball itself, its tactics move fast.

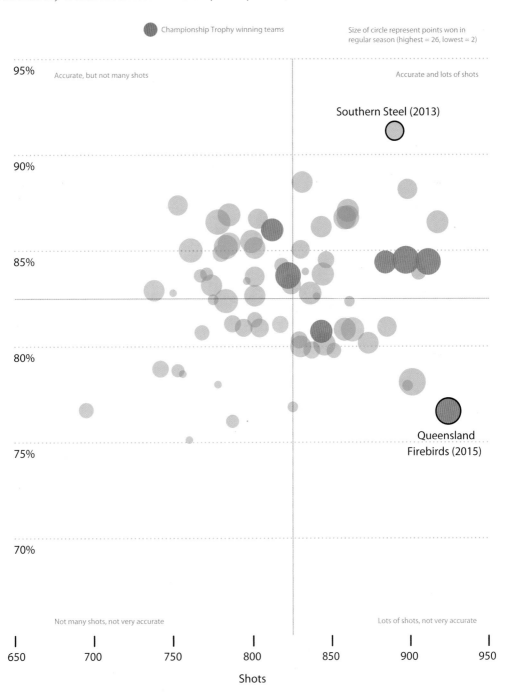

Netball: aim vs. volume

Total accuracy vs total shots for all ANZ Championship teams, 2009-15

Championship Trophy winning teams

Size of circle represent points won in regular season (highest = 26, lowest = 2)

95% Accurate, but not many shots

Accurate and lots of shots

Southern Steel (2013)

90%

85%

Accuracy %

80%

Queensland
Firebirds (2015)

75%

70%

Not many shots, not very accurate

Lots of shots, not very accurate

650   700   750   800   850   900   950

Shots

A world of world champions

There are world champions in some unlikely places, and in strange sports

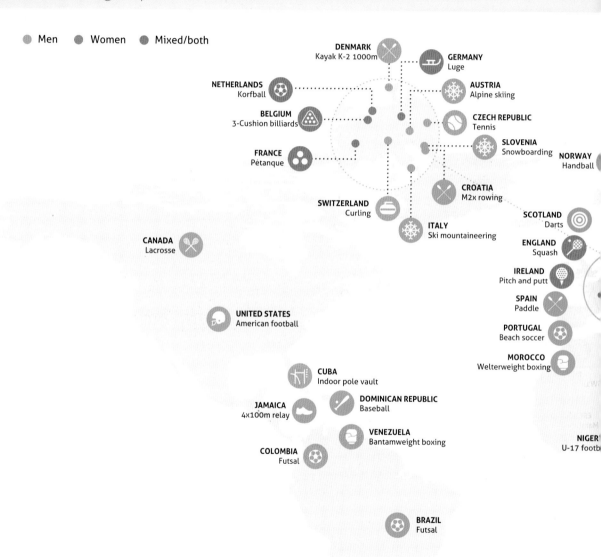

Men   Women   Mixed/both

DENMARK
Kayak K-2 1000m

GERMANY
Luge

NETHERLANDS
Korfball

AUSTRIA
Alpine skiing

CZECH REPUBLIC
Tennis

BELGIUM
3-Cushion billiards

SLOVENIA
Snowboarding

NORWAY
Handball

FRANCE
Pétanque

CROATIA
M2x rowing

SWITZERLAND
Curling

SCOTLAND
Darts

ITALY
Ski mountaineering

ENGLAND
Squash

CANADA
Lacrosse

IRELAND
Pitch and putt

SPAIN
Paddle

UNITED STATES
American football

PORTUGAL
Beach soccer

MOROCCO
Welterweight boxing

CUBA
Indoor pole vault

JAMAICA
4x100m relay

DOMINICAN REPUBLIC
Baseball

VENEZUELA
Bantamweight boxing

NIGER
U-17 footb

COLOMBIA
Futsal

BRAZIL
Futsal

ARGENTINA
Polo

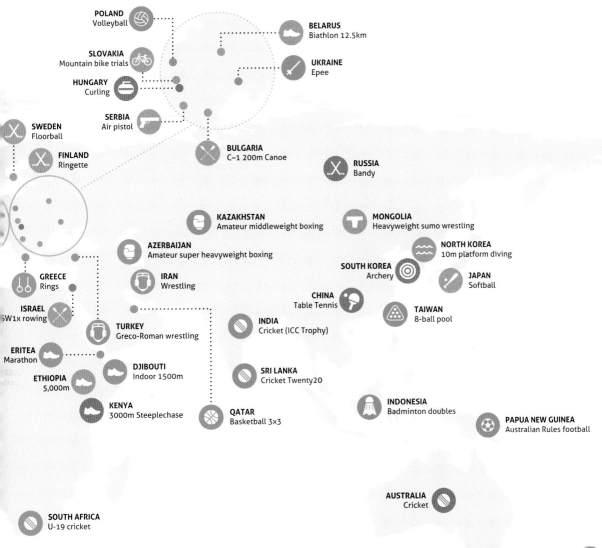

POLAND
Volleyball

BELARUS
Biathlon 12.5km

UKRAINE
Epee

SLOVAKIA
Mountain bike trials

HUNGARY
Curling

SWEDEN
Floorball

SERBIA
Air pistol

BULGARIA
C–1 200m Canoe

RUSSIA
Bandy

FINLAND
Ringette

KAZAKHSTAN
Amateur middleweight boxing

MONGOLIA
Heavyweight sumo wrestling

AZERBAIJAN
Amateur super heavyweight boxing

NORTH KOREA
10m platform diving

GREECE
Rings

IRAN
Wrestling

SOUTH KOREA
Archery

JAPAN
Softball

ISRAEL
W1x rowing

CHINA
Table Tennis

TURKEY
Greco-Roman wrestling

INDIA
Cricket (ICC Trophy)

TAIWAN
8-ball pool

ERITEA
Marathon

ETHIOPIA
5,000m

DJIBOUTI
Indoor 1500m

SRI LANKA
Cricket Twenty20

INDONESIA
Badminton doubles

KENYA
3000m Steeplechase

QATAR
Basketball 3x3

PAPUA NEW GUINEA
Australian Rules football

SOUTH AFRICA
U-19 cricket

AUSTRALIA
Cricket

NEW ZEALAND
Rugby union

How closely connected are the fortunes of drivers and constructors?

The winning F1 driver usually comes from the winning team – but not always

It's perhaps a paradox of Formula 1 that the public at large care about drivers, but the sport's insiders actually care more about teams.

Although drivers take the podium and the glory, the team focus makes sense when you consider that the constructor's championship essentially shows who has the best mechanics, designers, managers. It's a team sport, at heart. Drivers don't get far without a good car.

So despite occasional flare-ups between the two driving team mates, who are simultaneously competing against each other as well as trying to do best by the team, by the end of a Formula 1 season, there is an 85 per cent chance that the winning driver will be from the winning constructor.

That means that through F1 history, only 10 winning drivers have been from a (supposedly) lesser team. They are:

Mike Hawthorn (1958), Jackie Stewart (1973), James Hunt (1976), Nelson Piquet (1981 and 1983), Keke Rosberg (1982), Alain Prost (1986), Michael Schumacher (1994), Mika Häkkinen (1999) and Lewis Hamilton (2008).

What have those seasons got in common? They were all (bar one) incredibly close, with the championship won by just a handful of points.

This makes sense: in a closely-contested season, where the top two drivers are within a few points of each other, the constructor's title will be decided by the performance of the other drivers in a team.

So when we look at close seasons – those decided by five points or less – the chance of a driver being from the non-winning team rises, with nine out of 24. If we look at super-close seasons, where the difference was just two points, it is eight out of 13 – more than half the time.

As scoring systems have changed, with more points available, it also makes sense to look at the percentage difference between the top two drivers. If the second-placed driver is within 5 per cent of the winner, which has happened 16 times, it's a coin-toss: on eight occasions the winning driver has not been from the winning constructor.

The chart shows the percentage difference in points between the drivers at the end of each season and the total points scored by the winner. The new scoring system of the last few years is why some champions might win by a higher margin, but in percentage terms not be as far ahead as other drivers who have had dominant seasons.

So what is that strange outlier? In 1973, the championship was won by British driver Jackie Stewart, racing a Tyrrell car. Although Stewart won the driver's championship comfortably by 16 points with a total of 71, Lotus finished first with Brazilian Emerson Fittipaldi and Ronnie Peterson from Sweden in second and third places, on 55 and 52 points respectively. The way races were counted towards the constructor's championship that year meant that Lotus won by 10 points from Tyrrell.

Of course, theoretically, the constructors' championship is winnable by having two drivers get just over half of the total of the winning driver. Although it's mathematically possible, it never happens in practice. A season like 1973 may never come around again.

Team spirit

Winning constructors don't always make for winning drivers

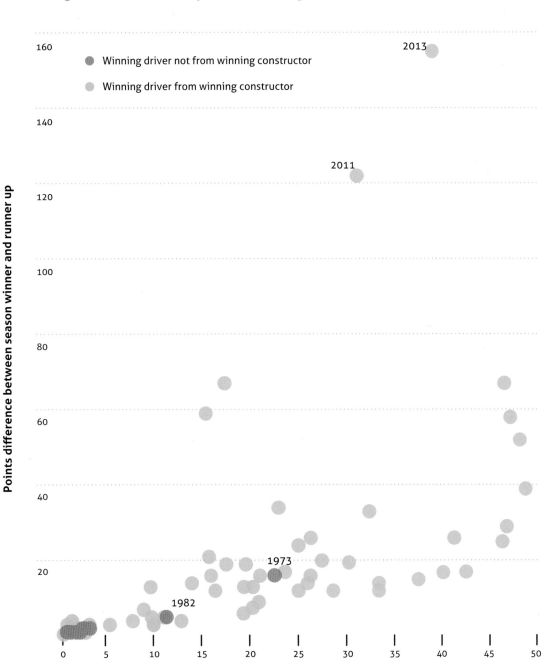

Points difference between season winner and runner up

- Winning driver not from winning constructor
- Winning driver from winning constructor

2013

2011

1973

1982

Percentage difference between season winner and runner up

The fastest drivers don't always come first

Can slow and steady really win the race?

Trying to work out who is the best Formula 1 driver of all time is a fruitless task. You can look at the absolute statistics, such as most wins or points – but that will skew the results to recent years, with more Grand Prix per season, and a higher-scoring points system. Over the years the cars have changed completely, the circuits have been improved, and a host of other smaller changes all add up to making comparisons between different generations almost impossible.

So let's ask a different question. Do fast drivers always win?

It seems an odd question on one level, when driving slowly isn't going to work. But being the quickest driver doesn't always mean being the winning driver.

F1 historical statistics going back to 1950 record both the winner of each race and also which driver set the fastest lap, as well as a few other details.

The first chart shows the F1 drivers who have had 10 or more Grand Prix victories in their career. We have then subtracted the wins where they also set the fastest lap, and then compared that to the races where they did set the fastest lap, but didn't win.

What emerges? Any driver placed above the line wins more times than they are quickest. Any driver below the line can drive quickly, but for whatever reason, doesn't cross the finish line first as often. The further a driver is to the top left of the chart, the more they are able to grind out wins. The closer a driver is to bottom right, the more they are flashy but with unfulfilled potential.

A few drivers stand out. Michael Schumacher is at the top right of the chart with lots of (relatively) slow wins, but also lots of fastest laps. The driver at the bottom right is Kimi Räikkönen, world champion in 2007, but far more of a speedster than a winner. And, perhaps surprisingly, the driver closest to the top-left corner is Ayrton Senna.

Frequently described as the most talented driver ever in F1, Senna was also a hard-headed pragmatist, who would do whatever it took to win. Perhaps he couldn't be bothered to show off and record the fastest lap if it didn't mean winning. Or maybe he just was more consistent than other drivers. Regardless, he didn't get many fastest laps when he won races either – of his 41 GP wins, he set the fastest lap in only 10, a comparatively very low proportion of just 24 per cent. (His rivals, Prost and Mansell, were both around 40 per cent, and Schumacher was over 50 per cent).

This doesn't take into account how many races a driver had in their career. What happens if we divide the two numbers by a driver's total races, to get a per-race figure?

The second chart shows those numbers. And again, Senna is top-left, and Räikkönen is out on his own at the bottom-right. Close to Senna are Jackie Stewart and the current drivers Lewis Hamilton and Sebastian Vettel. But instead of Schumacher, the great 1950s driver Juan Manuel Fangio is the driver at the top right.

How come the great Fangio has such a high proportion of fastest laps without a win? While he leads the way in F1 in terms of wins-per-race, does this undermine his credibility as the best driver ever?

This is a great example of how the scoring system changes the incentives for drivers. These days, the fastest lap in a Grand Prix is just a 'nice-to-have'. It doesn't mean anything in terms of points. However, that wasn't the case in the early days of F1 – the 1950 to 1959 seasons (Fangio's era) awarded one point for the driver recording the fastest lap. It was worth racing fast, especially given it was just eight points for the win, so a few extra points picked up through the season could make a difference.

Rather than these being show-off laps with no purpose, he was picking up points. In fact this is another statistic that shows how Fangio was, by many of the measures you care to use, the best driver ever.

Tortoise vs. hare: winning vs. fastest

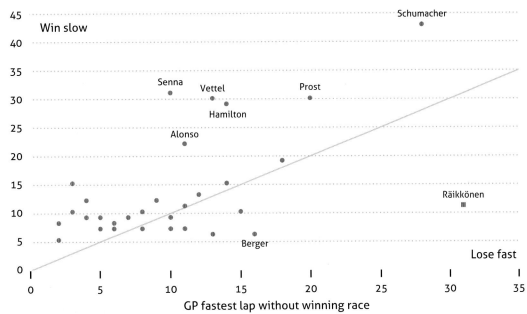

Winning slow vs. fastest laps, per race

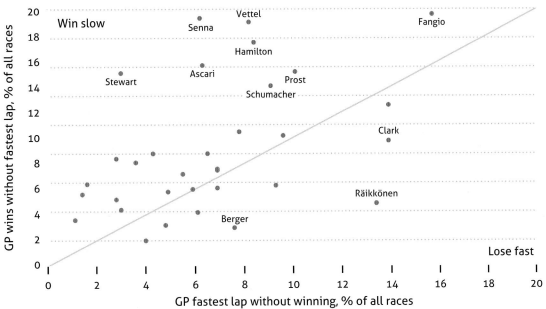

How much does pole position count in F1?

A clear start is less of an advantage than you might think

Pole position must clearly be an advantage in racing. No-one in front of you, a chance to get away from the pack of cars behind you, dodging and weaving about at the start. So it's no surprise that lots of Grand Prix winners start from pole; but has that changed over time?

As discussed on page 170-171, overtaking has had a renaissance in the last five years, after F1 had became something of a speedy procession. Surely, back in the dull days of the 1990s, there were lots of GP winners who simply led from start to finish?

Strangely not. Even in the 1990s, the rate of pole position winners per season varied from 19 per cent to 75 per cent. In other words, in a typical 16-race season, some years it might be as high as 12 pole winners, or as low as 3. Winning from pole doesn't fit with other things we know about F1 – average speed, or overtaking rates, for example.

One reason might be the make-up of the F1 calendar.

Some circuits are (almost) permanent fixtures in F1: Monaco, Silverstone, Monza, Spa. Others have come and gone, and on average there are a couple of changes each season to the calendar. In 61 years of F1, there have only been six seasons with the same circuits as the year before.

Leading from the front: pole wins and total races

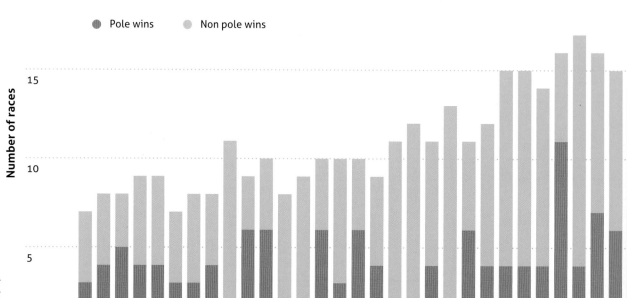

Which makes it hard to predict the pole effect – each circuit has its idiosyncrasies, so there is little discernible pattern. Take two of the regular circuits: Catalunya in Spain, and Monaco. Catalunya gets three pole winners in every four races; Monaco, with its short twisty circuit that makes it hard to overtake, has pole winners half the time. Yet Catalunya has more average lead changes per race (3.7 to 1.9) than Monaco.

Of course, as different circuits are added, the data becomes even more cluttered. Which is good – if pole position consistently determined the winner of a race, that would make F1 very predictable. Uncertainty is good for sport. And there are lots of things that affect the outcome of a Grand Prix: the weather, the actions of other drivers, the reliability of the car, good or bad pit stops.

What we can say with certainty is that there has never been a season with no winners from pole position. On average, over the history of F1, it's about 40 per cent of the time. Since 2010, it's been 50 per cent.

But even a lot of pole position winners don't mean boring seasons necessarily. In 1992, Nigel Mansell was in pole position 14 times out of 16, and converted 9 of those poles into wins. That was not an exciting season. But it was a lower pole-winner ratio than the season before, 1991, when 12 of 16 races were won by the first person on the grid. It might not have been a classic season (Ayrton Senna won with a race to spare), but it was certainly more entertaining than 1992.

In fact the season with the second-highest ratio of pole position winners was one of the closest ever and most exciting. In 1976, 11 of 16 races were won from pole. Yet James Hunt pipped Niki Lauda by just one point to the title in the last race of the season – and became the subject of the 2013 film *Rush*. Clearly, lots of pole-winners don't always lead to dull seasons. It's sharing the pole positions among different drivers that keeps things interesting.

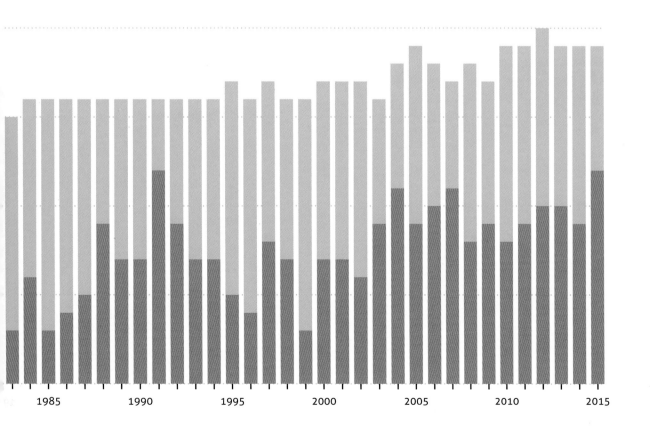

1985 1990 1995 2000 2005 2010 2015

How much difference does a scoring system make?

Formula 1 has changed the incentives, but has it changed the racing?

Talking about scoring systems might sound a bit dry, but don't be fooled. For any traditionalist sporting fan, they are sources of great debate and much anguish.

For anyone with a passing interest in game theory (the economic kind), they are a live incentive field experiment. Changes to a system often have unintended consequences, and Formula 1 has seen some of the most radical changes in how points are awarded in any major sport.

F1 has gone from the early days of awarding 8 points for a victory, to 25 points today. Along the way, there have been six different scoring systems, and various implementations of those systems where some, but not all, races counted to the season total. In effect, there have been 20 different scoring systems in 65 seasons.

If you then add in the different number of races, there have been 31 differently structured seasons out of 65.* That's a lot of tinkering. Too much, in fact. The high number of different seasons makes it hard to compare the systems effectively. In the real world of F1, the drivers and teams have been working it out as they have gone along, constantly adapting.

If we try to match up some of the supposed incentives with what actually happens on the track, some odd results get thrown up. The scoring system breaks down into five main groups, with the points awarded for first place, second and so on. They are:

8-6-4-3-2	1950 to 1959
9-6-4-3-2-1	1960 to 1990
10-6-4-3-2-1	1991 to 2002
10-8-6-5-4-3-2-1	2003 to 2009
25-18-15-12-10-8-6-4-2-1	2010 to present

I will refer to these as v8, v9, v10.1 v10.2 and v25.

If we take these scoring systems and show them as a percentage of the points available in a single race, we can compare how the points incentivise drivers.

It is clear from the first chart that the changes from v8 to v9 and then v10.1 gave greater emphasis to first place, as the percentage of total points in a race went up from 35 to 39 per cent. This was mainly at the expense of coming second or third.

You might expect this to encourage exciting racing, pushing drivers for the win. But other factors meant that the average number of lead changes under the v9 scoring system was lower. Cars got faster. Overtaking fell.

Then came the v10.2 system, which now awarded points for coming 7th and 8th, but also boosted places 3 to 6, mainly at the expense of coming first, which dropped from 38 per cent of the points to 26 per cent. With second place reduced slightly to 20 per cent of a race's total, the gap between 1st and 2nd was narrowed considerably. First was clearly less crucial.

Did this affect racing? You might expect it to encourage drivers to settle for second, with the difference being just two points. Why push for first and risk a crash? But something strange happened. Under v10.2, the number of average lead changes per race went up, not down. Under v10.1, there were an average 2.75 lead changes per race. Under v10.2, it was 3.93. That's quite a change. How come?

The answer is to look beyond the single race theory. F1 racing isn't just about single events. The number of races in a season is just as crucial.

Instead, if we look at the average maximum points available in each season under the different scoring systems and then show the points available in a race as a percentage of those available in the whole season, a different picture emerges.

Under v8, winning a race accounted for, on average, just under 19 per cent of the maximum points you could get that season. Under v25, with up to 20 races a season, that has dropped way down to 5.1 per cent. And in fact, rather than increasing and then falling as the single-race model suggests, the value of winning a Grand Prix has dropped with each system adopted. The increase in the average number of races counteracts part of the rebalancing of points.

Although v10.1 and 10.2 look very different on a race basis, when taken over the season, they look far more similar. Coming second was certainly better under v10.2, but by less than a percentage point. And the change in points for coming first from v10.1 to 10.2 was far less dramatic over a season than on an individual race basis. The drop is not from 38 to 26 per cent, but from 6 per cent to 5.7 per cent – a far less significant change than the single-race model suggests.

Instead, differences in lead changes can be attributed to a host of other rule revisions on areas such as tyres, downforce, engine size and so on, that combined to make the cars more equal.

Back in 2003, F1 introduced the v10.2 system to stop the season being won well before the end. In 2002, Michael Schumacher had the championship won with six races still remaining. It seemed a good idea to reduce the value of a win, relative to the other places, to maintain excitement to the end of the season.

Yet it didn't stop Schumacher winning the 2004 season with four races to go. Instead, by 2010, drivers and spectators felt that winning wasn't given enough prominence. Hence the move to 25 points, which on a per-race basis seemingly restores some of the weight to coming first.

In fact, there are fewer lead-changes per race on average under the v25 system than the v10.2, although not by much. Far from restoring a 'race-to-win' mentality, as administrators said at the time, the v10.2 points system was delivering excitement and a desire to win – as measured in lead changes – better than any system before or since.

Or, to be more accurate, other factors – luck, regulations, speeds – were affecting racing far more than any theoretical scoring incentives.

Judged on how many seasons are wrapped up early, it's clear that no points system can mitigate against a dominant driver: whether it is Schumacher in 2002 and 2004, or Vettel in 2011 and 2013, or Mansell in 1992.

Several retrospective studies have shown that very few F1 seasons would deliver different winners under different point systems. It also seems that, judged by how many times the lead changes hands, the points on offer don't make much difference either.

Perhaps the greatest incentive is simply to finish first. After all, it's the best way to win.

---

*Many seasons in F1 had a system where a driver could only count their best 11 races, or sometimes, there was something completely unique and only tried for one season, such as 1979: 15 races, counting the best 4 results from the first 7 and the best 4 from the last 8. No wonder it was only tried once.

Making a point

Scoring changes can be deceptive

Position of driver

First Second Third Fourth Fifth
Sixth Seventh Eighth Ninth Tenth

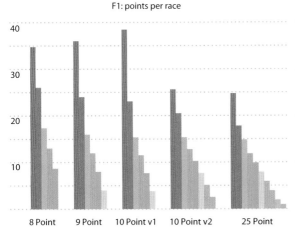

F1: points per race

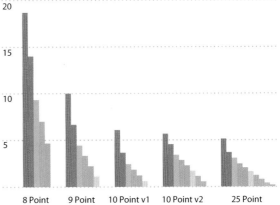

F1: Points per race in typical season

F1: balancing act
Can it be fast, exciting and safe?

F1 appears to face a long-standing trilemma. Think of the factory manager saying, 'you can have it cheap, fast or good – pick two' With F1 it's a case of picking two from fast, exciting, and safe.

Obviously, these things are relative – F1 is by definition a very fast sport, with inherent dangers. But across the sport's history, 'fast', 'exciting' and 'safe' can be measured in terms of average speed, number of overtaking manoeuvres, and deaths.

The trilemma seems to stack up. In the 1950s through to the 1970s, F1 was exciting: there were lots dramatic races and overtaking manoeuvres. It also got faster and faster. But it wasn't safe.*

Drivers were killed regularly: from 1950 to 1980, over

30 drivers died in crashes in F1-related activities (testing, qualifying and racing).

Of course, ever since the start of F1 there have been various safety measures introduced along the way, but the late-70s introduction of fireproof overalls and rollbars made a big difference to the drivers. In the last 20 years, there have been just four F1 deaths.

However, from the mid 1980s to the late 2000s, F1 became dull. This is the second phase of the trilemma: fast, safe, but not exciting.

Cars got fast, safety was increased, but the overtaking fell. A few teams and drivers became dominant. There were far fewer deaths. Since the great champion Ayrton Senna crashed into a wall and died at Imola in 1994, there has been

F1's trilemma: fast, exciting, safe?

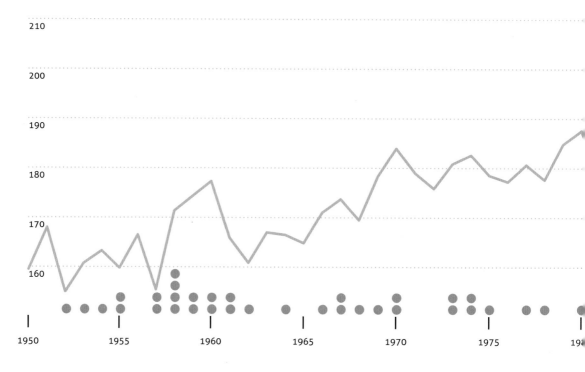

just one death, Jules Bianchi in 2015 (from an accident the previous season).

Yet fans were not happy with the level of entertainment. F1 was often described as a procession. The battle was fought through technology, with car design rather than driver skill the key factor.

From 2010 onwards, we have entered the third phase of the trilemma: safe, exciting, but not so fast. Rule and track changes meant that average F1 speeds fell from around 206 mph to under 190 mph – back to the speeds of the 1980s. The most important change was the introduction in 2011 of a drag reduction system, which gave an advantage to pursuing cars in certain areas on the track. Overtaking was back.**

So now F1 is exciting and safe but it's certainly less fast than it was before. Can it beat the trilemma, and increase speed again without sacrificing excitement and safety? The desire is there, with the proposed reintroduction of refuelling (banned since 2010, it allows cars to go faster) and changes to aerodynamics, announced in 2015. And although the decline in speed has been arrested, overtaking is now falling again.

However, the key to any dangerous sport's survival is making it safe enough to avoid controversy, and exciting enough to satisfy the fans. Interestingly, one of F1's proposed changes is to make engines noisier. Maybe just increasing the sound of speed will do.

---

*We have looked at deaths as a proxy for safety, rather than collisions. Strangely, the actual numbers of collisions and accidents per race fluctuates over time, unrelated to both speed and overtaking. Given that it only takes one bad move by one driver at the start of a race (the one moment all the cars are close together) to create mayhem and a mass accident, the data can be skewed by a couple of incidents.

**The available overtaking figures only go back to 1981. Source: Clip the Apex.

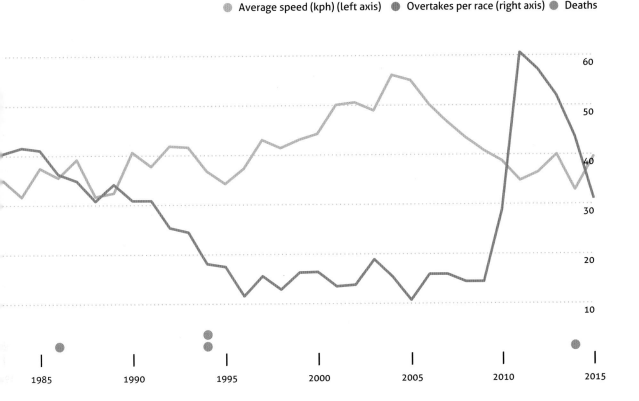

Average speed (kph) (left axis) ● Overtakes per race (right axis) ● Deaths

Statistics in sport: a timeline

It all adds up

All sports
Baseball

Book
Wisden Cricketers Almanack is first published. It is the longest-running sports annual in history

Research
English physiologist AV Hill defines VO2max, a measure of the maximum volume of oxygen that an athlete can use, which is still used to compare endurance today

1861   **1865**   **1920**   **1925**   **1955**

Book / Visionary
Bill James publishes his first Baseball Abstract, a collection of articles that attempts to answer baseball questions with statistics. Later, James defines the study as sabermetrics, based on the acronym of the Society of American Baseball Research (founded in 1971). James becomes the leading figure in baseball statistics and an influential thinker in sports worldwide

Innovation
Australian engineer James Horley develops the first real-time data telemetry for motor racing, revolutionising how pit teams can monitor cars during racing

Innovation
Cricinfo launches. The website, which is eventually bought by ESPN in 2007, is the first to feature ball-by-ball coverage on the internet in 1996. It also launches statsguru in 2000, a comprehensive search tool that gives readers free access to all sorts of statistics related to players and teams

1985   **1980**

Company
Prozone is founded. Derby county is first club to use the company's player tracking technology. Manchester United sign up in 1999

1990

1995   **2000**

Company
Sports data company Opta is founded, and quickly comes to attention via SkySports which uses the Opta Index to evaluate players' contribution (passes, shots, tackles etc) during matches

Appointment
Billy Beane becomes general manager of the Oakland Athletics, and starts a revolution in recruitment using data to find undervalued players. The A's reach the playoffs in four consecutive seasons

Innovation
Sir Clive Woodward has Prozone installed at Twickenham. Four years later, England win the World Cup

Debate
Alex Ferguson sells Jaap Stam to Lazio from Manchester United for £16.5m. Later, Ferguson describes the decision as a mistake, based on the wrong conclusion from Stam's tackling statistics

Innovation
Ball-tracking technology Hawk-Eye invented. It becomes an impartial third-party decision referee in tennis in 2006 and cricket in 2009

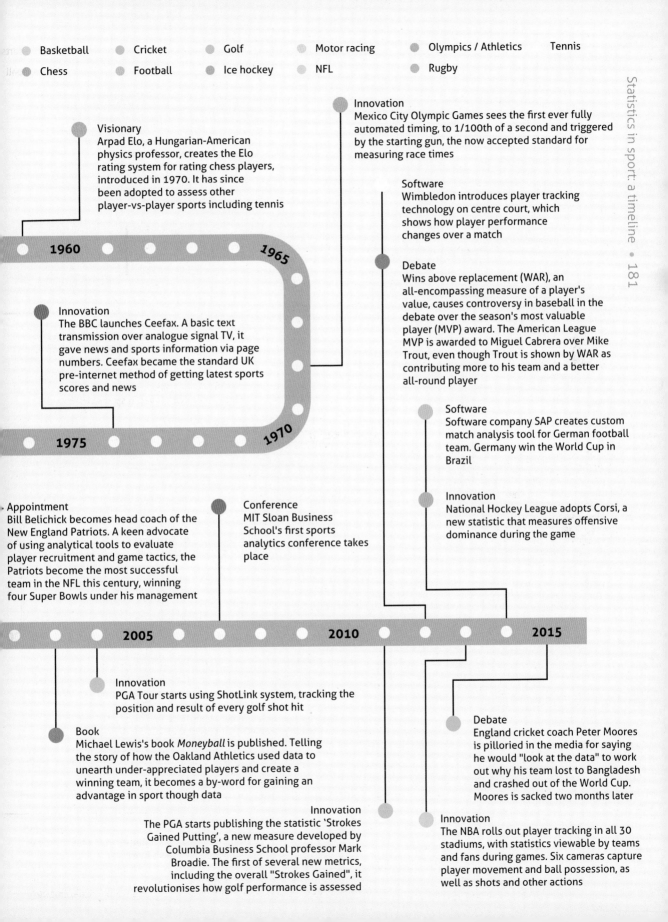

Basketball **Cricket** **Golf** **Motor racing** **Olympics / Athletics** **Tennis**
Chess **Football** **Ice hockey** **NFL** **Rugby**

Innovation
Mexico City Olympic Games sees the first ever fully automated timing, to 1/100th of a second and triggered by the starting gun, the now accepted standard for measuring race times

Visionary
Arpad Elo, a Hungarian-American physics professor, creates the Elo rating system for rating chess players, introduced in 1970. It has since been adopted to assess other player-vs-player sports including tennis

Software
Wimbledon introduces player tracking technology on centre court, which shows how player performance changes over a match

1960   **1965**

Debate
Wins above replacement (WAR), an all-encompassing measure of a player's value, causes controversy in baseball in the debate over the season's most valuable player (MVP) award. The American League MVP is awarded to Miguel Cabrera over Mike Trout, even though Trout is shown by WAR as contributing more to his team and a better all-round player

Innovation
The BBC launches Ceefax. A basic text transmission over analogue signal TV, it gave news and sports information via page numbers. Ceefax became the standard UK pre-internet method of getting latest sports scores and news

Software
Software company SAP creates custom match analysis tool for German football team. Germany win the World Cup in Brazil

1975   **1970**

Innovation
National Hockey League adopts Corsi, a new statistic that measures offensive dominance during the game

Appointment
Bill Belichick becomes head coach of the New England Patriots. A keen advocate of using analytical tools to evaluate player recruitment and game tactics, the Patriots become the most successful team in the NFL this century, winning four Super Bowls under his management

Conference
MIT Sloan Business School's first sports analytics conference takes place

2005   **2010**   **2015**

Innovation
PGA Tour starts using ShotLink system, tracking the position and result of every golf shot hit

Book
Michael Lewis's book *Moneyball* is published. Telling the story of how the Oakland Athletics used data to unearth under-appreciated players and create a winning team, it becomes a by-word for gaining an advantage in sport though data

Debate
England cricket coach Peter Moores is pilloried in the media for saying he would "look at the data" to work out why his team lost to Bangladesh and crashed out of the World Cup. Moores is sacked two months later

Innovation
The PGA starts publishing the statistic 'Strokes Gained Putting', a new measure developed by Columbia Business School professor Mark Broadie. The first of several new metrics, including the overall "Strokes Gained", it revolutionises how golf performance is assessed

Innovation
The NBA rolls out player tracking in all 30 stadiums, with statistics viewable by teams and fans during games. Six cameras capture player movement and ball possession, as well as shots and other actions

Index

Statistical glossary

Correlation

Correlation is when things 'go together', like bad weather and umbrella sales, or gin and tonic. Some things are correlated negatively, such as bad weather and ice-cream sales. More rain means fewer ice creams are sold.

If we take two groups of numbers, we can run a test on them and see how they score. If they are closely correlated, they will get a score approaching 1. Anything above 0.5 is considered a reasonable correlation. Anything getting close to zero shows the numbers are unconnected. The same goes for a score from -0.5 to -1, for things that are negatively correlated, (like our ice cream sales and rain).

But we have to be careful that we don't mistake correlation for causation. Perhaps tonic sales are up. And so are gin sales. More people are drinking gin and therefore want tonic with it, we might assume. But what if we missed something? Perhaps bitter lemon is also selling well, as is vodka. Are people drinking vodka and tonic, and gin with bitter lemon? We can't say for sure.

A statistical rule is: don't mistake correlation for causation. It's a good thing to bear in mind.

Standard deviation

When we have a set of numbers, it is useful to find out how spread out they are from their average. For that, we can use the standard deviation.

A high standard deviation of a group of numbers shows a wide spread; a low standard deviation score shows a closer grouping. This is a useful measure for how reliable we think polling data is, or showing how consistent a set of times is.

In case you are wondering, the formula for standard deviation is the square root of the average of the square of each number taken away from the average of set.

Rolling totals

Sometimes when we look at figures, whether month by month or year to year, the numbers leap up and down. It's hard to show what's going on. So to get an overall picture, we can use what is called a rolling average, or rolling total.

That means we look at the average number over, say four years. For sports with a four-yearly World Cup, that's often a good way of looking at numbers: the World Cup year may stand out for having an unusually-high number of matches, or some other aspect. Taking a four-year average means we can then look at the figures to see how things change over time in a more reliable way.

Economists often talk about data being seasonally-adjusted – people don't spend as much on say heating bills in summer as they do in winter, so you take a 12-month average to "smooth out" the data on consumer spending. This works in the same way.

Averages

There are three types of average.

In the following set of figures:

1, 1, 1, 1, 3, 5, 6, 7, 8, 9, 10

the **mode** is the most frequent number – in this case 1

The **median** – the one in the middle – is 5

And the **mean** (all the numbers added up and divided by how many there are) is 4.7. When people say **average**, this is the one they usually are referring to.

Sources

World Cup woes

Fifa

http://www.fifa.com/worldfootball/bigcount/registeredplayers.html
http://www.fifa.com/worldfootball/statisticsandrecords/tournaments/worldcup/teams/mostpartecipations.html

How many women football players are there anyway?

Fifa

http://resources.fifa.com/mm/document/fifafacts/bcoffsurv/bigcount.statspackage_7024.pdf
womensfootballsurvey2014_e_english.pdf

Uefa

2032784_DOWNLOAD.pdf
http://news.bbc.co.uk/sport1/hi/football/3402519.stm
http://fivethirtyeight.com/datalab/why-is-the-u-s-so-good-at-womens-soccer/

Maradona's 10 seconds of genius

Fifa

http://www.fifa.com/worldcup/matches/round=714/match=392/index.html#nosticky

YouTube

https://www.youtube.com/watch?v=icBYJMZur2U
https://www.youtube.com/watch?v=3z-qm-Sb_4s

The new champions, same as last time

Wikipedia

http://en.wikipedia.org/wiki/List_of_Italian_football_champions
http://en.wikipedia.org/wiki/List_of_German_football_champions
http://en.wikipedia.org/wiki/List_of_Spanish_football_champions
http://en.wikipedia.org/wiki/List_of_English_football_champions
http://en.wikipedia.org/wiki/List_of_European_Cup_and_UEFA_Champions_League_finals

Demography isn't destiny

Fifa

http://www.fifa.com/fifa-world-ranking/associations/association=chn/men/index.html
http://www.fifa.com/fifa-world-ranking/associations/association=ind/men/index.html

Should better teams get more penalties?

La Liga

http://www.football-lineups.com/tourn/La_Liga_2014-2015/Table/
and other years

Serie A

http://www.football-lineups.com/tourn/Serie_A_2014-2015/table/
and other years

EPL

http://www.football-lineups.com/tourn/FA_Premier_League_2014-2015/stats/penalties/
and other years

Bundesliga

http://www.football-lineups.com/tourn/Bundesliga_2014-2015/stats/penalties/
and other years

http://www.chelseafc.com/news/latest-news/2015/03/penalty-puzzle.html

European football's fallacy

La Liga

http://www.laliga.es/en/statistics-historical

Bundesliga

http://www.bundesliga.com/en/stats/table/

Premier League

http://www.premierleague.com/en-gb/matchday/league-table.html?season=1992-1993&month=MAY&timelineView=date&toDate=737074800000&tableView=CURRENT_STANDINGS

Serie A

http://www.worldfootball.net/schedule/ita-serie-a-1992-1993-spieltag/34/

Dangerous duopolies

The Rec.Sport.Soccer Statistics Foundation

http://www.rsssf.com/country.html

Uefa ranking for club competitions

http://www.uefa.com/memberassociations/uefarankings/country/index.html

US DoJ guidelines

http://www.justice.gov/atr/horizontal-merger-guidelines-08192010

The most complicated tournament in the world

Uefa

www.uefa.com/MultimediaFiles/Download/Regulations/uefaorg/Regulations/02/23/69/59/2236959_DOWNLOAD.pdf

The sporting jamboree and corruption

The Guardian

http://www.theguardian.com/sport/2014/oct/02/oslo-withdrawal-winter-olympics-2022-ioc

Transparency International

http://www.transparency.org/cpi2014/results

Reuters

http://uk.reuters.com/article/2013/04/24/us-soccer-fifa-idUSBRE93N18F20130424

Home comforts

IOC

http://www.olympic.org/Documents/Reference_documents_Factsheets/Sports_on_the_programme_of_the_Olympic_Winter_Games.pdf

Faster, higher, easier?

Wikipedia

http://en.wikipedia.org/wiki/Athletics_at_the_2012_Summer_Olympics_%E2%80%93_Men's_100_metres#Heat_3_3
http://en.wikipedia.org/wiki/Athletics_at_the_2012_Summer_Olympics_%E2%80%93_Qualification#Men.27s_200.C2.A0m
http://en.wikipedia.org/wiki/2012_United_States_Olympic_Trials_(track_and_field)
http://en.wikipedia.org/wiki/Olympic_sports

IAAF

http://www.iaaf.org/records/toplists/sprints/100-metres/outdoor/men/senior/2012

USA track and Field

http://www.usatf.org/Events---Calendar/2012/U-S--Olympic-Team-Trials-TF/Results.aspx

Some countries flourish at both Summer and Winter Olympics

IOC
http://www.olympic.org/olympic-studies-centre/documents-re-ports-studies-publications

Sports Reference
http://www.sports-reference.com/olympics/summer/
http://www.sports-reference.com/olympics/winter/

Wikipedia
https://en.wikipedia.org/wiki/South_Korea_at_the_Olympics

Drugs, hosts and obsessions

Economic Prediction of Medal Wins at the 2014 Winter Olympics, Andreff and Andreff
https://www.researchgate.net/publication/227473343_Economic_Prediction_of_Medal_Wins_at_the_2014_Winter_Olympics

BBC
http://www.bbc.co.uk/sport/winter-olympics/2014/medals/countries
http://www.bbc.co.uk/news/world-europe-26276980

SB Nation
http://www.sbnation.com/2014/2/26/5405120/winter-olym-pics-2014-netherlands-speed-skating-medals-dominance

Olympics host medals
Modelling home advantage in the Summer Olympic Games, N J Balm-er, Alan Michael Nevill, A M Williams
https://www.researchgate.net/publication/10673396_Modelling_home_advantage_in_the_Summer_Olympic_Games
Olympic.org
Wikipedia
Sports Reference
Own analysis

One billion? Don't believe the hype

http://www.independent.co.uk/sport/football/news-and-com-ment/why-fifas-claim-of-one-billion-tv-viewers-was-a-quarter-right-5332287.html
http://www.nielsen.com/us/en/insights/news/2008/beijing-olym-pics-draw-largest-ever-global-tv-audience.html
http://www.sportingintelligence.com/2012/07/26/london-2012-be-ware-billions-bollocks-ceremony-to-be-huge-tv-hit-but-not-that-huge260701/
http://qz.com/171174/putting-the-global-perspective-into-the-su-perbowls-massive-tv-audience/
http://www.bbc.co.uk/sport/30326825
http://www.dawn.com/news/1167592

Google trends
https://www.google.co.uk/trends/explore#q=%2F-m%2F01l10v%2C%20%2Fm%2F06brs%2C%20%2Fm%2F021vk%2C%20%2Fm%2F09rt3%2C%20%2Fm%2F086px&cmpt=q&tz=Etc%2FGMT

Major league, major problem

Baseball Reference
http://www.baseball-reference.com/leagues/MLB/1916-misc.shtml
http://www.baseball-reference.com/leagues/MLB/2015-misc.shtml
http://www.beyondtheboxscore.com/2014/2/10/5390172/ma-jor-league-attendance-trends-1950-2013
https://en.wikipedia.org/wiki/World_Series_television_ratings

Can Moneyball be measured?

http://courses.cs.washington.edu/courses/cse140/13wi/projects/mirae-report.pdf
https://eh.net/encyclopedia/the-economic-history-of-ma-jor-league-baseball/
http://www.baseball-reference.com/bullpen/2000_Chicago_White_Sox
http://roadsidephotos.sabr.org/baseball/data.htm
http://www.thecubdom.com/Papers/FCI/fancostindex_printerfriendly.html

https://web.archive.org/web/20061230052302/http://www.team-marketing.com/fci.cfm?page=fci_mlb2006.cfm
http://time.com/3774279/mlb-ticket-price/

The winners and losers of baseball's 1994 strike

http://www.baseball-almanac.com/players/birthplace.php
http://www.si.com/mlb/2014/08/12/1994-strike-bud-selig-orel-her-shiser
http://www.usatoday.com/story/sports/mlb/2014/08/11/1994-mlb-strike/13912279/
http://www.ibtimes.com/huge-salaries-poverty-stricken-country-eco-nomics-baseball-dominican-republic-1546993
http://americasquarterly.org/node/2745

No bases for old men?

http://www.baseball-reference.com/leagues/MLB/pitch.shtml
http://www.baseball-reference.com/leagues/MLB/bat.shtml

The Balco effect

http://www.alltime-athletics.com/m_100ok.htm
http://www.baseball-reference.com/leagues/MLB/bat.shtml

The end of world records?

http://www.iaaf.org/records/by-category/world-records
http://www.iaaf.org/news/iaaf-news/moscow-2013-statistics-book-available-to-down
http://www.pgatour.com/hole-in-one/news/2014/06/11/season.html
http://www.alltime-athletics.com/w_1500ok.htm

Why Bolt could do better

http://www.alltime-athletics.com/m_100ok.htm
http://www.alltime-athletics.com/m_200ok.htm

Flo-Jo's legacy

http://www.alltime-athletics.com/w_200ok.htm
http://www.alltime-athletics.com/w_100ok.htm

Is the 2-hour marathon in reach?

IAAF
http://www.theguardian.com/lifeandstyle/the-running-blog/2014/jan/31/sub-two-hour-marathon-possible
http://rw.runnersworld.com/sub-2/
http://regressing.deadspin.com/in-search-of-a-two-hour-mara-thon-1641691951
http://www.nytimes.com/2014/09/30/upshot/forecasting-the-fall-of-the-two-hour-marathon.html?ref=sports&abt=0002&abg=1&_r=0
http://sportsscientists.com/2010/08/the-sub-2-hour-marathon-who-and-when/

What's small, round and can't be thrown over 70m any more?

Alltime athletics
http://www.alltime-athletics.com/wdiscok.htm

Switching flags

https://en.wikipedia.org/wiki/List_of_nationality_transfers_in_sport

An anatomy of the greatest try

http://www.bbc.co.uk/programmes/p00ndc2n
http://www.bbc.co.uk/sport/0/rugby-union/21205137
http://www.barbarianfc.co.uk/results-fixtures/1972-1973/

How important is home advantage?

ESPN Scrum
http://stats.espnscrum.com/statsguru/rugby/stats/index.html?-class=1;filter=advanced;orderby=date;page=2;size=200;tem-plate=results;trophy=27;type=team;view=results
http://stats.espnscrum.com/statsguru/rugby/stats/index.html?-class=1;filter=advanced;orderby=date;page=3;size=200;span-min1=01+Jan+1996;spanval1=span;template=results;trophy=2;-

type=team;view=results
http://stats.espnscrum.com/statsguru/rugby/stats/index.html?-class=1;filter=advanced;orderby=date;page=3;size=200;span-min1=27+Sep+1999;spanval1=span;template=results;trophy=2;-type=team;view=match
http://stats.espnscrum.com/statsguru/rugby/stats/index.html?-class=1;filter=advanced;orderby=date;page=3;size=200;span-min1=27+Sep+1999;spanval1=span;template=results;trophy=2;-type=team;view=match
http://www.rugbyfootballhistory.com/scoring.htm

When the lions play, who wins?
http://en.wikipedia.org/wiki/Six_Nations_Championship
ESPN
http://stats.espnscrum.com/statsguru/rugby/stats/index.html
http://stats.espnscrum.com/statsguru/rugby/stats/index.html?class=1;filter=advanced;opposition=1;opposition=10;op-position=2;opposition=20;opposition=3;opposition=4;oppo-sition=5;opposition=6;opposition=8;opposition=9;orderby=-date;page=19;size=200;spanmin1=01+Jan+1947;spanval1=span;team=1;team=10;team=2;team=20;team=3;team=4;team=5;team=6;team=8;team=9;template=results;type=team;view=match

Are rugby players bigger than ever?
ESPN

The battle for third
http://www.independent.co.uk/sport/football/news-and-com-ment/why-fifas-claim-of-one-billion-tv-viewers-was-a-quarter-right-5332287.html
http://www.nielsen.com/us/en/insights/news/2008/beijing-olym-pics-draw-largest-ever-global-tv-audience.html
http://www.sportingintelligence.com/2012/07/26/london-2012-be-ware-billions-bollocks-ceremony-to-be-huge-tv-hit-but-not-that-huge260701/
http://qz.com/171174/putting-the-global-perspective-into-the-su-perbowls-massive-tv-audience/
http://www.bbc.co.uk/sport/30326825
http://www.dawn.com/news/1167592
Google trends
https://www.google.co.uk/trends/explore#q=%2F-m%2F01l10v%2C%20%2Fm%2F06brs%2C%20%2Fm%2F021vk%2C%20%2Fm%2F09rt3%2C%20%2Fm%2F086px&cmpt=q&tz=Etc%2FGMT
Web Search interest: UEFA European Championship; Rugby World Cup; The Cricket World Cup; Commonwealth Games; Winter Olympic Games
Worldwide; 2004 - present

What 2-point conversions tell us about the NFL and risk
http://www.pro-football-reference.com/years/NFL/scoring.htm#sea-son_totals

The rise of the quarterback
http://www.nfl.com/player/danmarino/2501869/gamelogs?sea-son=1984
http://www.pro-football-reference.com/years/ALL/team_stats.htm
http://www.pro-football-reference.com/leaders/pass_yds_single_sea-son.htm
http://www.pro-football-reference.com/leaders/pass_yds_year_by_year.htm
http://www.pro-football-reference.com/leaders/pass_td_year_by_year.htm
http://www.boston.com/sports/football/patriots/arti-cles/2011/10/16/why_the_passing_game_is_spiraling_out_of_control/?page=full
http://www.footballperspective.com/the-ebb-and-flow-of-the-nfl-passing-game-since-1932/
http://bleacherreport.com/articles/1089533-how-the-rise-of-the-passing-game-in-the-nfl-has-taken-over-the-league

NFL lessons in geography
http://www.nfl.com/news/story/0ap3000000401027/article/nfl-envi-sions-londonbased-team-in-2022-notes-across-league
http://en.espn.co.uk/more/sport/story/354795.html
http://mmqb.si.com/2014/10/22/nfl-international-series-global-ex-pansion
http://www.ibtimes.com/nfl-2014-league-continues-see-benefit-reg-ular-season-games-london-1705735
http://fivethirtyeight.com/features/the-nfl-should-expand-to-london-but-first-canada-mexico-and-la/

The case for college football
http://www.ncaa.org/championships/statistics/ncaa-football-attend-ance
http://fs.ncaa.org/Docs/stats/football_records/Attendance/2014.pdf
http://www.cbssports.com/collegefootball/writer/jon-solo-mon/25329428/college-football-attendance-making-rare-in-crease-early-in-2015
http://www.sportingintelligence.com/finance-biz/business-intelli-gence/global-attendances/
http://espn.go.com/nfl/attendance/_/year/2014
http://espn.go.com/nfl/attendance/_/year/2015
http://www.stadiumsofprofootball.com/comparisons.htm
https://en.wikipedia.org/wiki/List_of_American_football_stadiums_by_capacity
NFL.com
https://en.wikipedia.org/wiki/List_of_attendance_figures_at_domes-tic_professional_sports_leagues

An anatomy of the Drive
Pro Football Hall of Fame
http://www.profootballhof.com/history/decades/1980s/the_drive.aspx
YouTube
https://www.youtube.com/watch?v=GWGx_P9KPZg
ESPN
http://espn.go.com/nfl/playoffs/2013/story/_/id/10370685/nfl-best-championship-clutch-drives

Why stadiums are getting smaller
http://stadiumdb.com/constructions
http://www.worldstadiums.com/stadium_menu/past_future/past_sta-diums.shtml
http://www.worldstadiumdatabase.com/
http://www.thesportmarket.biz/charts/stadiumstats/top50stadiums/ranktop.html
http://www.totalprosports.com/2011/10/27/11-most-expensive-sta-diums-in-the-world/
http://www.bbc.co.uk/news/world-asia-35158004

All-rounders need time to prove their worth
ESPN Cricinfo
http://stats.espncricinfo.com/ci/content/records/282786.html

Recounting the centuries
ESPN Cricinfo
100s
http://stats.espncricinfo.com/ci/content/records/227046.html
200s
http://stats.espncricinfo.com/ci/content/records/230344.html
300s
http://stats.espncricinfo.com/ci/content/records/282944.html

England and the ODI fallacy
ESPN Cricinfo
http://stats.espncricinfo.com/ci/engine/stats/index.html?class=2;or-derby=year;team=1;template=results;type=team;view=year
http://stats.espncricinfo.com/ci/engine/stats/index.html?class=2;or-derby=year;team=2;template=results;type=team;view=year

http://stats.espncricinfo.com/ci/engine/stats/index.html?class=2;or-derby=year;team=3;template=results;type=team;view=year
http://stats.espncricinfo.com/ci/engine/stats/index.html?class=2;or-derby=year;team=4;template=results;type=team;view=year
http://stats.espncricinfo.com/ci/engine/stats/index.html?class=2;or-derby=year;team=5;template=results;type=team;view=year
http://stats.espncricinfo.com/ci/engine/stats/index.html?class=2;or-derby=year;team=6;template=results;type=team;view=year
http://stats.espncricinfo.com/ci/engine/stats/index.html?class=2;or-derby=year;team=7;template=results;type=team;view=year
http://stats.espncricinfo.com/ci/engine/stats/index.html?class=2;or-derby=year;team=8;template=results;type=team;view=year

How far can cricket's one-day scoring record go?

ESPN Cricinfo
http://stats.espncricinfo.com/ci/content/records/284258.html

Drugs in sport: a timeline

http://usatoday30.usatoday.com/sports/2007-02-28-timeline-drugs-athlete_x.htm
http://sportsanddrugs.procon.org/view.timeline.php?time-lineID=000017

Millionaires everywhere

PGA
http://www.pgatour.com/stats/stat.109.html
http://www.pgatour.com/champions/stats/stat.109.2008.html

http://www.europeantour.com/europeantour/racetodubai/pastsea-son/index.html#bLBBWfHJsS5S9m3y.97

The short stuff matters

PGA
http://www.pgatour.com/stats/stat.101.html

The magic number is 18

http://www.augusta.com/masters/players/bios/Jordan_Spieth.shtml
http://www.augusta.com/masters/historic/players/hbh/1997_hbh184.shtml

Golf's great global shift?

http://en.wikipedia.org/wiki/Official_World_Golf_Ranking
http://www.owgr.com/about?tabID={BBE32113-EBCB-4AD1-82AA-E3FE9741E2D9}
http://www.telegraph.co.uk/sport/golf/rydercup/11129498/America-will-soon-give-up-on-the-Ryder-Cup-altogether-if-its-teams-keep-on-being-hammered.html
http://www.telegraph.co.uk/sport/golf/rydercup/11127049/Ryder-Cup-2014-How-the-two-captains-Paul-McGinley-and-Tom-Watson-compared.html
https://en.wikipedia.org/wiki/Men%27s_major_golf_championships
http://www.rydercup.com/europe/history/2014-ryder-cup-past-re-sults
http://www.golf.com/ap-news/europe-losing-top-players-pga-tour

Where have all the American golfers gone?

NGF
http://docslide.us/documents/state-of-the-industry-10-golf-demand-participants-mm-source-ngf-golf-participation-study-1990-is-a-calculated-average-of-1989-and-1991-long-term.html
http://www.statista.com/statistics/275308/number-of-registered-golf-players-in-europe/
http://www.ega-golf.ch/050000/050200.asp
http://www.ega-golf.ch/050000/050100.asp
http://www.golfdigest.com/story/number-of-golfers-steady-more
http://ngfdashboard.clubnewsmaker.org/aotj5q6ddeyhv3uqgk3u-w7?email=true&a=3&p=2184375&t=43825
http://www.bloomberg.com/news/articles/2014-01-16/golf-course-closings-outpace-openings-for-eighth-straight-year
http://www.golfdigest.com/story/china-continues-to-wage-its-war-on-golf-makes-joining-golf-clubs-illegal

Snooker just keeps on getting better

http://www.cuetracker.net/Statistics
http://www.cuetracker.net/Statistics/Centuries/Tourna-ment-Frames-per-Century-Rate/Season/1980-1981
http://www.cuetracker.net/Statistics/Centuries/Most-Maximums-147s-Made/Season/1980-1981

How tennis is killing doubles

Wimbledon annuals
http://www.independent.co.uk/sport/tennis-wimbledon-raising-the-stakes-profits-and-prizemoney-for-the-worlds-most-famous-tour-nament-break-new-ground-john-roberts-reports-1372878.html
http://longislandtennismagazine.com/article5837/jensen-zone-dou-bles-game-about-flatline
http://www.usopen.org/en_US/about/history/prizemoney.html

From Murray to Perry: seventeen degrees of separation

Slam History, ATP, Wimbledon
http://www.slamhistory.com/en/tennis/gbbb/John_Newcombe

The tennis paradox: win the battle, lose the war

https://github.com/JeffSackmann/tennis_wta
https://github.com/JeffSackmann/tennis_atp

Women tennis players are still losing the battle for equal pay

http://www.tennis.com/earnings/WTA/
http://www.tennis.com/earnings/ATP/
ATP
atp_media_guide_2015.pdf
WTA
http://www.wtatennis.com/press-center

Giant killing in tennis

ATP, Jess Sackmann via Github
Incentives in Best of N Contests: Quasi-Simpson's Paradox in Tennis
International Journal of Performance Analysis in Sport, Volume 13, Number 3, December 2013, pp. 790-802(13)
http://www.ingentaconnect.com/content/uwic/ujpa/2013/00000013/00000003/art00019
http://www.theatlantic.com/entertainment/archive/2014/01/what-every-pro-tennis-player-does-better-than-roger-federer/283007/

parkrun

http://www.parkrun.com/
http://www.parkrun.org.uk/bushy/results/eventhistory/
http://www.parkrun.org.uk/aboutus/organisation/
http://www.usatf.org/events/2012/OlympicTrials-TF/entry/qualifying-Standards.asp

Is the Tour de France getting easier?

http://www.letour.fr/
http://www.letour.fr/le-tour/2014/docs/Historique-VERSION_INTE-GRALE-fr.pdf
http://www.letour.fr/le-tour/2015/docs/Historique-VERSION_INTE-GRALE-fr.pdf
http://bikeraceinfo.com/tdf/tdfstats.html

Burden of proof

Dopeology
http://www.dopeology.org/statistics/incidents/

The hour mark: back from the dead

https://en.wikipedia.org/wiki/Hour_record
http://www.uci.ch/track/news/article/timeline-modern-uci-hour-re-cord/

http://www.uci.ch/mm/Document/News/NewsGeneral/16/60/64/
Historiquedesrecords_HommesElite_02.06.2015_Neutral.pdf
http://www.telegraph.co.uk/men/active/11440145/Jack-Bobridge-At-
tempting-cyclings-hour-world-record-is-the-closest-you-can-come-
to-death-without-dying.html

In defence of Sally Robbins
http://theboatraces.org/results

Race data
http://www.worldrowing.com/events/2004-olympic-games/wom-
ens-eight/final/
http://library.la84.org/6oic/OfficialReports/2004/Results/Rowing.pdf
http://www.independent.co.uk/news/world/australasia/olympic-
rower-who-stopped-takes-part-in-australian-parade-and-gets-a-
slap-6161480.html
http://www.sports-reference.com/olympics/athletes/ro/sally-rob-
bins-1.html
http://news.bbc.co.uk/sport1/hi/olympics_2004/rowing/3597914.
stm
http://www.theage.com.au/articles/2004/08/23/1093246448746.
html
http://www.telegraph.co.uk/news/worldnews/australiaandthepa-
cific/australia/1472133/Public-humiliation-for-Australian-row-
er-who-gave-up-in-Athens-final.html
http://www.abc.net.au/lateline/content/2004/s1183326.htm

That's the price you pay
http://www.statista.com/statistics/274826/tickets-available-and-
sold-at-the-olympic-summer-games/
http://www.statista.com/chart/2377/costs-of-hosting-the-world-cup/
http://news.bbc.co.uk/1/hi/world/africa/8718696.stm
http://www.reuters.com/article/soccer-world-brazil-idUSL2E8F-
2GG820120403
https://www2.uni-hamburg.de/onTEAM/grafik/1098966615/Plessi,-
Maennig-(2007),-World-Cup-2010.pdf
http://www.theguardian.com/sport/2012/nov/13/olympic-tickets-lo-
cog-coe-deighton
http://www.olympic.org/Documents/Reports/EN/en_report_1428.pdf
http://www.olympic.org/documents/reports/en/en_report_900.pdf
http://www.olympic.org/Documents/Reports/EN/en_report_274.pdf
http://www.fifa.com/about-fifa/official-documents/governance/index.
html#financialReports
http://www.statista.com/statistics/265030/summer-olympics-broad-
casting-revenue/
http://www.olympic.org/Documents/IOC_Marketing/OLYMPIC_MAR-
KETING_FACT_%20FILE_2014.pdf

Kangaroos rarely get caught
http://www.rugbyleagueproject.org/matches/Custom/LS0tLS0tLS0tL-
S0tLXktLS0tLS0tLQ==?page=4

The not-so-Super League
http://rugby.statbunker.com/competitions/HomeAttendance?comp_
id=469
http://www.premiershiprugby.com/stats/attendance.php?inclu-
deref=18099&season=2014#18099#CXO4jExg8muriE3V.97
http://www.rugbyleagueproject.org/competitions/super-league/
seasons.html

Michael Jordan is the greatest player ever
http://www.basketball-reference.com/players/j/jamesle01.html
http://www.basketball-reference.com/players/j/jordami01.html
http://www.basketball-reference.com/leaders/

Losing: a winning strategy
http://www.basketball-reference.com/leagues/NBA_wins.html
http://espn.go.com/nba/story/_/id/12000240/lakers-owner-jean-
ie-buss-says-tanking-unforgivable
http://uk.businessinsider.com/sam-hinkie-explains-sixers-tanking-
plan-2015-2?r=US&IR=T
http://sports.yahoo.com/nba/blog/ball_dont_lie/post/Did-the-Cava-
liers-tank-to-help-draft-LeBron-Jame

http://espn.go.com/blog/truehoop/post/_/id/40055/does-tanking-
even-work
https://en.wikipedia.org/wiki/List_of_National_Basketball_Associa-
tion_seasons
http://espn.go.com/boston/nba/story/_/id/9434966/for-boston-celt-
ics-tanking-likely-necessary-never-easy
http://www.cheatsheet.com/google-news/5-nba-teams-that-success-
fully-tanked-for-the-draft.html/?a=viewall
http://bleacherreport.com/articles/2310039-nba-insider-the-fine-art-
of-tanking-and-of-how-the-commissioner-talks-about-it

Just a long-range shooting contest
http://www.basketball-reference.com/leagues/NBA_stats.html
http://www.sportingcharts.com/articles/nba/the-rise-of-the-three-3-
pointer.aspx
http://espn.go.com/nba/playoffs/2015/story/_/id/12993098/nba-35-
year-war-3-pointer
http://bleacherreport.com/articles/1715367-how-the-3-point-shot-
has-revolutionized-the-nba

Bodysuit technology and swimming world records
http://www.sportsrecords.co.uk/swimming/
http://www.fina.org/content/swimming-world-ranking
http://www.fina.org/sites/default/files/wr_50m_oct_1_2015_rede-
signed.pdf
http://usatoday30.usatoday.com/sports/olympics/london/swimming/
story/2012-08-03/swimming-world-records-fall-despite-ban-su-
per-tech-suits/56734742/1

Front crawl is a bit of a misnomer
http://www.sportsrecords.co.uk/swimming/
http://www.fina.org/content/swimming-world-ranking
http://www.fina.org/sites/default/files/wr_50m_oct_1_2015_rede-
signed.pdf

Secretariat stands alone
Belmont
http://www.belmontstakes.com/past-winners.aspx

Kentucky
http://www.kentuckyderby.com/history/kentucky-derby-winners

Preakness
http://www.horse-races.net/library/preak-winners.htm
http://www.vancouversun.com/sports/This+history+-
June+1973/6764080/story.html
http://fivethirtyeight.com/features/where-have-you-gone-secretariat/

Are horses getting quicker?
http://rsbl.royalsocietypublishing.org/content/11/6/20150310.
article-info
http://rsbl.royalsocietypublishing.org/content/11/6/20150310.
figures-only
Patrick Sharman, Alastair J. Wilson, 2015
http://www.alltime-athletics.com/m_100ok.htm

Canada has lost the NHL
League history
https://en.wikipedia.org/wiki/List_of_Stanley_Cup_champions
https://en.wikipedia.org/wiki/History_of_organizational_chang-
es_in_the_NHL
http://www.legendsofhockey.net/LegendsOfHockey/jsp/SilverwareT-
rophyWinners.jsp?tro=STC

Nationality data
http://www.quanthockey.com/nhl/nationality-totals/nhl-players-
2013-14-stats.html
and other years

Ratings and background
http://fivethirtyeight.com/features/why-cant-canada-win-the-stanley-
cup/
http://www.hollywoodreporter.com/news/why-nhl-tv-ratings-

are-806708
http://www.cbc.ca/sports/hockey/nhl/game-7-smashes-hockey-night-in-canada-record-1.1015705?cmp=fb-cbcsports

National Sports Act
http://laws-lois.justice.gc.ca/eng/acts/n-16.7/page-1.html

Fighting on Ice
http://www.hockeyfights.com/leaders/teams/1/reg1960
http://www.hockeyfights.com/leaders/teams/1/reg2015
http://www.nhl.com/stats/team?season=20032004&game-Type=2&viewName=summary
http://www.nhl.com/nhl/en/v3/ext/rules/2014-2015-rulebook.pdf
http://edition.cnn.com/2011/SPORT/09/01/nhl.enforcers.deaths/
http://www.nytimes.com/2011/09/02/sports/hockey/deaths-of-three-nhl-players-raises-a-deadly-riddle.html?_r=0

We're going to need a bigger net
http://www.hockey-reference.com/leagues/stats.html
http://www.wsj.com/articles/are-nhl-goalies-too-big-to-fail-1450657491
http://www.thehockeynews.com/blog/is-it-time-for-the-nhl-to-go-to-bigger-nets-mike-babcock-still-thinks-so/
http://www.thestar.com/sports/leafs/2015/11/11/players-coaches-divided-on-idea-of-bigger-hockey-nets.html

Big screen, small success
IMDb
The Numbers
http://www.the-numbers.com/movie/records/#world
Worldwide Boxoffice
http://www.worldwideboxoffice.com/

Going down, sooner
BoxRec
http://boxrec.com/titles

The greatest boxer ever you (probably) never heard of
BoxRec
http://boxrec.com/records

Paid to play, or paid to smile?
Forbes
http://www.forbes.com/athletes/list/#tab:overall

It's all downhill from here
FIS (International Ski Federation)
http://data.fis-ski.com/

Where Andorra is bigger than China, for now
http://data.fis-ski.com/global-links/statistics/statistical-graphs.html?statstype=licence

Up in the air
ANZ Championship
http://www.anz-championship.com/Statistics?Year=2015
http://www.abc.net.au/news/2015-06-21/queensland-firebirds-net-ball-gretel-tippett/6562032
https://www.youtube.com/watch?v=qxiALDb0zmQ
http://mc.championdata.com/anz_championship/index.html?matchid=95640107

A world of world champions
http://en.wikipedia.org/wiki/List_of_world_sports_championships

How closely connected are the fortunes of drivers and constructors?
http://f1-facts.com/stats/seasons/different-Winners
http://f1-facts.com/stats/seasons/closest-championships
http://f1.wikia.com/wiki/Points

The fastest drivers don't always come first
Ergast.com
DriverDB
Race-database.com

How much does pole position count in F1?
http://race-database.com/season/season.php

How much difference does a scoring system make?
http://f1.wikia.com/wiki/Points

F1: balancing act
http://ergast.com/mrd/
http://ergast.com/api/f1/status/4/results?limit=1000
http://ergast.com/api/f1/status/3/results?limit=1000
DriverDB
https://www.driverdb.com/championships/standings/formula-1/2014/
Race-database.com
http://race-database.com/season/season.php
Clip the Apex
http://cliptheapex.com/overtaking/seasons/2015/
http://cliptheapex.com/overtaking/circuits
http://cliptheapex.com/overtaking/seasons
http://formula1.markwessel.com/blog/
https://en.wikipedia.org/wiki/List_of_Formula_One_fatalities

Statistics in sport: a timeline
FT, Misc, BBC, ESPN, Opta, Prozone, Wired, PGA, Sloan, WSJ, Calgary Herald

Acknowledgements

This book has taken a very long time. So the following people have been very patient. Thank you.

Family / support

Joanna Cooke
Matilda, Frankie, Annie and Jimmy
Mary and Graeme Minto
Judy and Fra Cooke
Lucy and Catherine, Crail and Colin
Carmen Funuyet Cáceres

Making it happen

Sophie Lambert (Conville & Walsh)
Charlotte Croft (Bloomsbury)
Sarah Connelly (Bloomsbury)

Book title

Vanessa Cooke

Main writing locations

The British Library (thanks to the wonderful staff)
Literary Cafe, Tufnell Park
Next door (thanks to the neighbours)

Inspiration and ideas

Ash Anderson
Tom Burgis
John Burn-Murdoch
James Coleman
Joanna Cooke
James Fauset
Jim Freeman
Tom Hill
Michael Hunter
Jamie Inman
Ian Jacobsberg
Sam Kay
Sarah Laitner
Richard Marsh
Malcolm Millar
Sam Raphael
James Ratzer
Moreis Scobie
Alan Smith
Leo Stemp
Alex Stirling
Richard Thomas
Andy Weinberg